How to buy high-fidelity

Bring quality audio into your home

Copyright © 2020 Riccardo Ruggiu
All rights reserved

*Many thanks to all the people
I've meet during these years: music teachers,
sound technicians/engineers, audiophiles, musicians and DJs,
because without their precious teachings and
suggestions, this book wouldn't
have been made.*

Acknowledgements

Thanks to Miriam, my family and friends, for having always been present.

*Music is everywhere, it accompanies us throughout life.
A voice singing on the street, the sound of a musical instrument, the soundtrack of a film, the sounds of nature.
With its rhythm, it marks the time of millions of people around the world.
Music is life!*

R.Ruggiu

Limit of Liability/Disclaimer of Warranty

The author make no representations or warranties with respect to the accuracy or completeness of the contents of this work and specifically disclaim all warranties, including without limitation warranties of fitness for a particular purpose.
The advice and strategies contained herein may not be suitable for every situation.
If professional assistance is required, the services of a competent professional person should be sought.
The author shall not be liable for damages arising hereform.
The fact that an organisation is referred to in this work as a citation and/or a potential source of further information does not mean that the author endorses the information the organization may provide or recommendations it may make.
Further, readers should be aware that the author shall not be liable for damages arising hereform.

Introduction

Dear Stereo Fan,
this book is meant to building solid foundations and bringing home high fidelity.
Maybe you're thinking to buy a new sound system (or a vintage one... why not?) but you are confused by the information the seller has given you and by the conflicting opinions you read on the internet.
With the return of the vinyl records, the turntable reappeared as a source also in the latest hi-fi generation systems enjoying great success among passionates.
Unfortunately, however, it's not enough to buy a good turntable to enjoy listening to vinyl records; in fact, many owners of turntables, even very expensive, cannot get close to the full potential of the component.
A meticulous setup is necessary, in order to make the turntable play at its best and make it able to extract as much musical infomations as possible from the groove.
The book deals wih the issues of amplification, the various kinds of loudspeakers and wiring.
It also includes numerous tips about the sound system positioning and about the listening environment.

This book doesn't pretend to answer all the questions on the subject, but it's intended to be a useful guide to make your sound system work at its best.

After finishing reading this book, you'll be able to enjoy the highest quality possible from your system.

I just have to wish you happy reading and... happy listening!

Contents

Chapter I   Audio equipment categories   pag. 1

A little bit of sound theory   pag. 4

Better an "all in one" system or better to choose separate components?   pag. 8

Chapter II   The sources   pag.14

The cd player   pag.15

The turntable   pag.20

- Direct drive and belt drive   pag.23
- Tonearms   pag.25
- MM or MC cartridge?   pag.29
- Conical or elliptical stylus?   pag .31
- High compliance and low compliance   pag.32

- Turntable placement   pag.34
- Turntable setup and tune up   pag.36
  1. Leveling the turntable   ''
  2. Cartridge alignment   pag.40
  3. Tonearm balancing   pag.46
  4. Antiskating adjustment   pag.50
  5. Fine speed controls   pag.55
- Conclusions   pag.58
- About the load capacity (pF), impedance (kOhm) and output capacity (mV)   pag.59
- The network audio player   pag.63

Chapter III   The Amplifier   pag.65

About amplification and "Clipping"   pag.71

Amplifier classes in a nutshell   pag.78

- Class A   ''
- Class AB   pag.79
- Class D   pag.80

Chapter IV	Loudspeakers	pag. 82
-	Frequency response	pag. 83
-	Sensitivity	pag. 88
-	Impedance	pag. 91
-	Power	pag. 93
-	Floor loudspeakers	pag. 96
-	Bookshelf loudspeakers	pag. 97
	Types of cabinets	pag. 99
-	Horn loudspeakers	''
-	Infinite baffle	pag.100
-	Passive radiator (or A.B.R.)	pag.101
-	Finite baffle	pag.102
-	Bass reflex	pag.103
-	Isobaric system	pag.104
-	Electrostatic loudspeakers	pag.106
	Wireless? No, thanks!	pag.108
	Full range, 2 ways, 3 ways or more?	pag.109

Chapter V	Interconnect cables	pag.112
	- RCA cables	pag.113
	- XLR cables (or Cannon)	pag.116
	- Coaxial digital audio cable	pag.121
	- Optical digital audio cable	pag.122
	Which to choose: optical or coaxial?	pag.124
	Power cables	pag.125
Chapter VI	Listening room and sound system positioning	pag.130

Chapter I
Audio equipment categories

We can distinguish electronics for playing music in three categories:
- *Low-Fi*
- *My-Fi*
- *Hi-Fi* and *Hi-End*

The first category (*Low-Fi*) includes devices that give back low quality when playing music; for example, mp3 players played via docking stations or boom boxes, mini speakers of the laptop or your smartphone's speakers.

Let's move on to the second category "*My-Fi*".
Many people perceives high fidelity according to the music played by their system (whatever it is).
Many users who have never heard a musical instrument, or a live concert, claim that their system faithfully reproduces music because they are addicted to the sound of their system (even if probably does not correctly reproduce the sound of many musical instruments and / or the voices of singers).

Once during a live concert, at the end of a reggae song the singer said:"You listened the rhythm right, because he was on the drums! From Jamaica, Mr.....!".
This was to make it clear that the rhythm of the drums was perfect, because played by a musician from the homeland of reggae!
So I introduce the third category (_Hi-Fi_ and _Hi-End_), which aims to reproduce sounds and music in the most faithful way possible to reality; from the original sound of an intrument to the voice of the singers.
To know if a system sounds "right", you should have heard the musical instruments live, so that you can make a comparison with the system you are listening to.
All this, regardless of the cost of the electronics but based exclusively on the best performance of all the components of the audio chain in the listening environment.
Regarding Hi-End, this category includes the most expensive audio products ever and which are designed without compromise, by manufacturers specialized in a particolar sector (for example, they design only turntables, only amplifiers, only speakers and not an entire audio chain from the source to the loudspeakers).

However, there are also cases of medium-level electronics manufacturers, who have given examples of the skills achieved, designing electronics that have become part of the Hi-End sector).

Regardless of the price, Hi-End also includes products that represent the highest technological / construction level relating to the construction period.

For example, a Hi-End amplifier produced in the 80s or 90s, maintains high quality and construction standards that are difficult to overcome or equal even by the best current products; just think about a supercar of the 80s that reaches 200 mph, is difficult to overcome even today, despite the technological progress and materials evolution.

Remember that 10 different audio chains will make you hear the same CD in 10 different ways, but perhaps only one will make you listen the right sound!

A little bit of sound theory

Sound is produced by regular vibrations of elastic bodies.
The properties of sound are: height, intensity, timbre and duration.
The height of a sound is its being more acute, more serious or the same compared to another.
The height of the sound depends on the number of vibrations that a vibrating body performs in a second.
The International reference unit is the hertz indicated with the Hz symbol.
The human ear does not percerive all sounds, but only a minimal part between 20 Hz and 20.000 Hz (in ideal conditions within a laboratory, it has been verified that the human ear can hear sounds between 12 Hz and 28.000 Hz); over time, this hearing ability decreases.
All frequencies below 16 Hz are called infrasound, those above 20.000 Hz are called utrasound.

The table below represents the audible frequencies by the human ear:

Frequency Range	Frequency Values
Sub-bass	20 to 60 Hz
Bass	60 to 250 Hz
Low midrange	250 to 500 Hz
Midrange	500 Hz to 2 kHz
Upper midrange	2 to 4 kHz
Presence	4 to 6 kHz
Brilliance	6 to 20 kHz

The intensity of the sound is the force with which a sound can be heard, which can be: strong, very strong, soft, very soft, even if the height doesn't change.

The causes of the intensity of the sound are: the amplitude of the surface of the vibrating body and the resonance phenomena, and the distance between the vibrating body and the listener.

The intensity of a sound can be measured in decibels (dB).

Starting with silence (0dB) here are some examples; rustling of leaves 10 dB, normal conversation at 1 meter 40-60 dB, intense traffic at 10 meters 80-90 dB, pneumatic hammer at 1 meter and disco around 100 dB, jet engine at 100 meters and rock concert about 110 dB.

The dangerous range for hearing is between 90 and 110 dB and above.

If you want, you can download some applications for your smartphone and have fun checking the intensity of sounds around you.

Regarding the timbre, is intended the quality that allows us to distinguish which source is producing a sound; is the feature that allows you to distinguish sounds of equal pitch and intensity: for example a pipe organ from a trumpet, or the voice of an individual from another. But also a "C major" chord from a "C minor" chord.

Duration, is the length of time the sound persists; it can be a long or short sound.

It is measured in seconds; the shorter the sound, the harder it is recognize it.

Better an "all in one" system or better to choose separate components?

The choice of the system is probably the most difficult moment of the journey, to get a home sound reproduction as faithful as possible to the sound scenario with which the album (CD or LP) was recorded in the studio, rather than live in a theater or a stadium.

In this case, the available budget plays a fundamental role and this is the main variable that guides the purchase of a stereo system.

It is possible to follow two different paths: rely on a complete kit (which usually includes a CD player, a tuner, an amplifier and a pair of speakers) sold in specialized shops, but also in large retailers, or you can choose each component individually, in specialized electronics stores.

The choice of the sound system, depends on the needs of the user.

Of course, cheaper solutions will provide a lower quality result.

If you decide to purchase separate components, selecting the best product from each manufacturer, remember that the quality choice must follow the figure of an upside down pyramid.

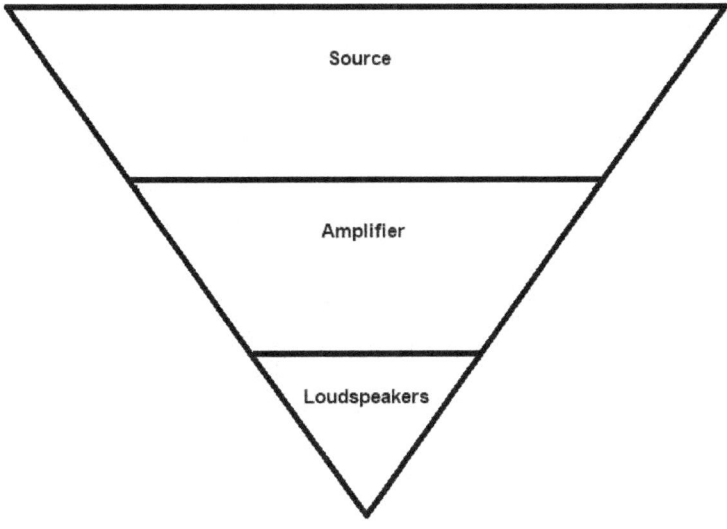

In fact, the source (or sources) must have better quality than the rest of the audio chain; therefore it will be necessary to allocate the largest part of the budget.

Immediately after, we find the amplifier (which can be integrated or separated in two electronics, preamplifier and power amplifier).

And last but not least, the speakers.

The qualitative scheme of the inverted pyramid is very important because if I spend little money on a source, a higher amount of money for the amplifier, and even more money to buy the speakers, the signal would come out of the speakers highlighting the performance limits of the source and the amplifier, transforming all the defects detected during the passage through the audio chain into sound.

Instead, following the qualitative scheme of the inverted pyramid in the figure seen above, the signal will arrive to the speakers already filtered and free of unwanted anomalies collected along the passage in the electronics audio chain.

Usually the "mini hi-fi" stereo system (or "compact stereo"), are designed to satisfy the needs of consumers who approach the world of Hi-Fi for the first time and have a limited budget and a space that does not allow to host large systems.

With a few hundred dollars, you can start getting familiar with these devices without spending crazy amounts of money.

Aesthetically they "do not disturb" the furniture too much, being quite compact and with small-sized speakers.

By choosing this type of system, there will be no compatibility issues between the various components and there will be only one remote control to manage the entire system.

However, even if you choose a kind of system of this type, you must not forget that in order to allow the system to express itself in the best way, it's <u>always</u> necessary to position the speakers correctly in the room dedicated to listening.

The choice of an excellent quality system cannot always bring the desired results, as it could be oversized compared to the environment and give an unsatisfactory audio performance.

IMPORTANT: if you decide to purchase a system with separate components, rely on top quality cables with gold-plated contacts. Even if the cables are already in the boxes, avoid them.

For optimal listening, cables specially designed for high performance are available.

Chapter II
The sources

As suggested earlier, the source (or sources if more than one) should have higher quality than the rest of the audio chain; this is to avoid carrying shortcomings and defects of the audio electronics along the signal path.

We can divide the sources into three types: CD players, turntables, and network audio players (or if you are planning to complete a vintage system, tuners).

The choice must also be made according to needs, because if you have a collection of 30 CDs and 1000 vinyl records at home, it will be useless to concentrate on the purchase of a CD player.

Conversely, if you love the sound of vinyl, it will be useless to go and see what the CD players market offers.

Once you have established what your priorities are for purchasing the source, you can start checking its characteristics and models offered by various companies.

The CD player

Regarding *CD players*, we can divide them into two categories; integrated cd players, or in two separated frames, CD player + Dac (digital analog converter). A bit like amplifiers, that are divided into two categories (integrated amplifier or preamplifier + power amplifier) as we'll see later on.
Nowadays there are high-level integrated CD players, which do not disfigure at all, compared to the two-frame solution.
To obtain "absolute performances", the solution of two separate frames is always better (provided that they are produced by the same brand); this is very important, since they are designed to work perfectly together in order to avoid issues of electrical and sound compatibility.
The advantages of the "two frames" are the separate power supplies (one for the mechanics and one for the signal conversion) that allow the electronics to work better (the best CD player, built in a single frame, adopt two separate tranformers to power the digital circuits

separately from those for the analog section), and then there is another important thing; in fact the vibrations produced by the mechanics are kept away from the chassis that houses the conversion section.

The disadvantages of the "two frames" are: the greater dimensions, compared to the single frame of the integrated CD and the use of more cables to be used for the connections and last but certainly not least, the cost.

Some audiophiles, to improve the sound of their integrated CD player, use the digital output to connect it to an external digital / analog converter in order to increase the signal quality; this procedure increases the criticality of the combination, because the dac that is added, was not created to "work in tandem" with the integrated CD and therefore it would be useful to try different dac models and evaluate which offers the best sound.

To ensure greater flexibility of use, it is important that the device is compatible with the largest number of supports. Whether you decide to buy a new CD player or a vintage one, check that it is able to read both the cd-r (recordable) and the cd-rw (rewritable); super audio CD players (SACD) have a higher cost, both as regards electronics (intended for audiences of Hi-End audiophiles), and as regards the support recorded with this technology.

The SACD is able to provide a more defined soud image with exceptional realism; the frequency response reaches up to 100 kHz and the dynamic range up to 120 dB against the 20 kHz and 100 dB of the best CD players.

Among the accessory functions, the output for connecting headphones with a dedicated amplifier is useful.

Some CD players are compatible with USB sticks for playing Mp3 files (or even WAV, WMA, etc.), or for converting the CD into Mp3 files without having to use the computer.

Through the USB port, they can interact with iPods and Mp3 players.

The remote control is also very useful, to facilitate remote interaction wih the CD player (if you buy a used CD player, make sure it is included).

Regarding the connections, the cheaper CD players, have the RCA analog audio outputs and the coaxial digital output; higher level CD players, have also optical digital output and balanced XLR (or Cannon) output.

Another quality index factor (and not only for CD players but for all electronics in general) is weight.

In fact, the use of anti-resonance materials, designed to stiffen the electronics structure both to isolate it from external vibrations, and from those produced by the mechanics itself, tend to increase weight making it massive.

An inexpensive integrated CD player will weigh between 3 and 5 kg. (mainly with plastic structure), while a high-end CD player will weigh between 14 kg. and 16 or more (with mainly metal structure).

As for the technical specifications, which play an important role for the good performance of the electronic device, I indicate below some of the higher values among vintage and new CD players:

- Frequency response between 2Hz and 20.000Hz (up to 50/100 kHz SACD). Although this frequency spectrum is practically identical for most CD players, some readers have a lower low range and start from 20 Hz to 20.000 Hz.
- Dynamic range: 100 dB (120 dB SACD)
- Signal to noise ratio: 117 dB
- Channel separation: 110 dB
- Total harmonic distortion: 0,0025%

Generally speaking, any cd player that has or does close to these technical characteristics, it is certainly of a high standard.

However, the last word is always up to listening since each hi-fi component has a different way of "extracting music" from a CD; therefore electronics that may look similar on paper, could play music in a completely different way.

The turntable

Believe it or not, very few people around the world listened to a vinyl record correctly played by their turntable.

A little bit like it happened with the audio compact cassettes, which many hated, because were recorded incorrectly or reproduced with poor cassette players, ignoring their true potential and not enjoying their sound.

When the compact disc was introduced on the market, many got rid of vinyls and audio cassettes, enchanted by the "perfection" of the CDs.

Unlike the turntable (wich needs to be "tuned" after purchase), and the cassette deck (which must be well adjusted to record audio cassettes at its best), the CD player needs no tune up and once you insert the CD, you can enjoy a perfect reproduction.

After this necessary premise, we begin the journey into the mythical world of turntables.

There are various types of turntables, for all needs and budgets.

I'll describe the various features and types of turntables, so that you know how to deal with the vinyl jungle according to your needs.

It starts with the cheapest, are already complete with all the components (base, arm and cartridge) and then going up with the price only the base and the arm (cartridge to buy separately), or otherwise only the base (with arm and cartridge to buy separately).

There are two categories of bases: suspended and non suspended one (as example, suspended on elastic springs).

Having tried both solutions, I can say that both are valid; probably, with the floating solution, the vibrations of the room in which the turntable is located (or if you prefer the building in which you live) are better damped.

So if you live in an apartment where every time the door of the building is closed wildly, you hear the floor and the walls vibrate, probably the suspended base wil be the most suitable for you (because the non suspended one, would transmit vibrations to the arm and head, generating vinyl tracking errors and adding noises not present on the vinyl to the signal sent to the amplifier).

If instead you live in an independent house, where the floors are not too elastic (to vibrate when walking or jumping) and in a quiet residential area, then you can also choose a non suspended one.

I'll explain later on how to position the turntable effectively.

Direct drive and belt drive

There are two kinds of rotation system: direct drive and belt drive.
There are several excellent examples from both categories from the new or second-hand market.
Think about the use you are going to do of your turntable, and also about the budget at your disposal.
If you haven't much money, choose a belt drive turntable, because it offers good performances at an affordable price; this kind of turntables, manage to dampen vibrations through the belt wich unfortunately deteriorates in the long run, causing general decay of the turntable and a loss of the regular rotation speed of the platter (be careful if you have in mind to buy some used one).
Direct drive turntables have a higher cost, because, having the motor directly connected to the platter, they use high quality motors that have less vibrations.

Direct drive turntables, where used in disco and radio broadcasting both, because they were able to make lightning-fast starts of the plate, but also fast enough to stop.

They needed less maintenance, and without belt they guaranteed a perfect rotation speed of the platter, even after years.

Although the most hardened and nostalgic audiophiles turn up their noses when it comes to direct drive, it must be admitted that the turntables that use it sound very good.

Tonearms

Regarding the tonearm, there are four types: straight tonearm, S-tonearm, J-tonearm and tangential.

Most of the tonearms in the market are the straight ones, since they have a lower production cost than the other three and have good performances, even if they are often low in mass and cannot "handle" any type of cartridge.

Straight tonearm

Then there are the S-shaped ones, which are more robust and able to handle more demanding cartridges in terms of weight and sound; on average they are more expensive than straight tonearms.

S-Tonearm

Like the S-Tonearm, the J-Tonearms also are able to trace the groove of the vinyl with greater accuracy and can be combined with low compliance cartridges.

J-Tonearm

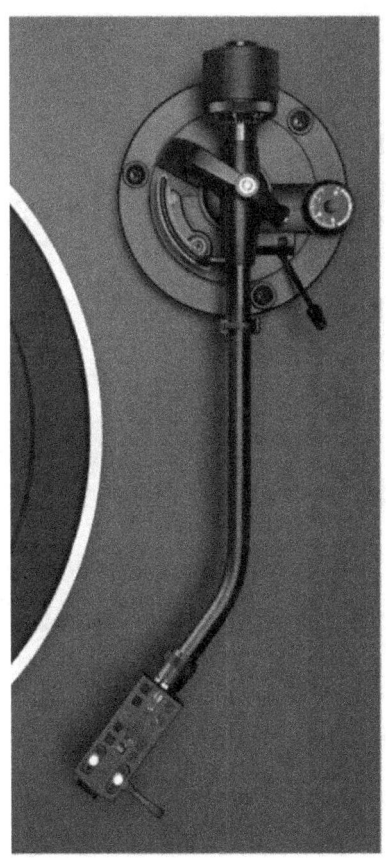

And to conclude, the tangential tonearm, usually assembled inside turntables with futuristic features and very beautiful to look at (even if they are not at the top together with the best existing turntables).

Tangential tonearm

Of course, turntables with tangential tonearm will make a good impression in a living room, however it remains at a lower level than the other three.

MM or MC cartridge?

And here we are, at the cartridge! Can be of two types: **MM** (moving magnet) or **MC** (moving coil).

MM cartridges are generally cheaper, can be interfaced with most existing amplifiers and a new replacement stylus can be purchased once it has finished its lifecycle.

MC cartridges are the most expensive and also those that achieve the best performances (as long as they are combined with the correct tonearm and base, as well as the amplifier / preamplifier that must have the input capable of supporting the low voltage signal output of an MC cartridge).

Once the life cycle of the MC head has ended, it must be completely replaced, or it can be restyled by a specialized technician (since on MC, the cantilever is an integral part of the cartridge itself).

It should be noted that both MM and MC must be mounted and aligned correctly in order to be able to extract the greatest amount of musical information from the vinyl record.

You could also buy the best MC cartridge in world, but if you mount it badly...it will also sound bad.

If you are a beginner, an MM cartridge will probably be fine for you, as it can get very close to an MC in terms of quality, while maintaining a moderate cost and greater ease of interfacing with most of the preamplifiers and integrated amplifiers currently on the market and vintage.

Conical stylus or elliptical stylus?

The conical stylus is the one with the simplest shape, forgives dimming errors and is present on the least expensive cartridges.

The elliptical stylus, due to its shape, better tracks the grooves of the disc and has a greater ability to extract musical informations (especially regarding high frequencies).

It presents a greater installation criticality, as it does not tolerate dimming errors and usually has a higher cost.

In general, the conical points are mounted on low quality cartridges, although there may be exceptions; the elliptical stylus are mounted mainly on medium/high quality cartridges.

High compliance and low compliance

To put out the best performances, the tonearm and the cartridge should be matched.

Not all cartridges will work with all tonearms and you must check the specifications of the cartridge and the tonearm.

High compliance cartridges (usually the cheapest ones) support a lower tracking force weight and having a lower rigidity, require more antiskating intervention.

Vice versa, low compliance cartridges support a greater tracking force weight and precisely because of their rigidity, they are able to work even without antiskating; in fact, some tonearms that combine with low compliance MC cartridges, don't have an antiskating adjustment knob.

Be careful not to overdo with the tracking force weight of the cartridge, because despite being able to tolerate it, you will increase the wear and tear of your record collection by shortening its life; for home use (unless you are a DJ and use the turntable to scratch) a reading weight between 2 and 3 grams is recommended even if the cartridge you like can handle a tracking force over 3 grams.

In professional use, such as radio broadcasting or DJ, antiskating was disabled in the past, to avoid errors in reading a song (ex.reading the end of the previous song, or song already started due to "landing" of the stylus in the wrong groove).

However, in the domestic environment, to have the best sound performance and at the same time avoid damaging the grooves of the vinyl and also the stylus (especially with high compliance) deforming it after a few hours of listening, it's recommended to use and optimize the tune up of the antiskating (if present).

Turntable placement

Avoid placing the turntable near sources of magnetic flux, such as on or near amplifier, or equipment containing one or more transformers such as the picture below:

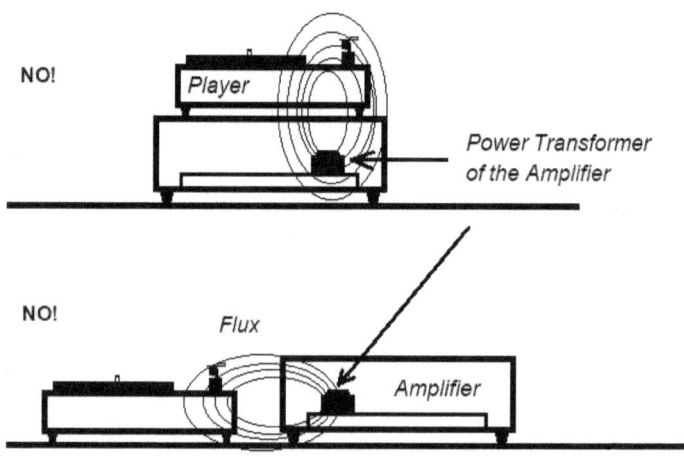

Since the turntable is sensitive to vibrations, do not place it directly in front of a speaker, on top of a speaker, or in contact with a wall / wall bracket, etc. These may induce howling. Allow for adequate ventilation to dissipate motor heat, which may become significant during long listening sessions.

Turntable setup and tune up

Now that we have examined the various types of turntables, tonearms and cartridges, it's time to discuss the topic about how to proceed to the various settings.
Unfortunately, most of the turntables found in the homes of vinyl music enthusiasts around the world aren't tune up before playing records; this implies that the turntable, although of excellent quality and high cost, is not put in the condition to be able to provide the best performances since the necessary adjustments have not been made.

1. Leveling the turntable

The first adjustment to be made is leveling the table or the piece of furniture on which we will place the turntable.
It's important that the base is heavy, rigid and "deaf" (that is, it does not receive vibrations transmitted by the environment, rerouting them towards the turntable causing acoustic feedback).

With a bubble level (also a cheap one), we must check that the table where the turntable will rest on is perfectly leveled.

Once the turntable rests, remove the rubber from the platter and check with the bubble level that is perfectly leveled as shown in the photos below:

Bubble level adjustments

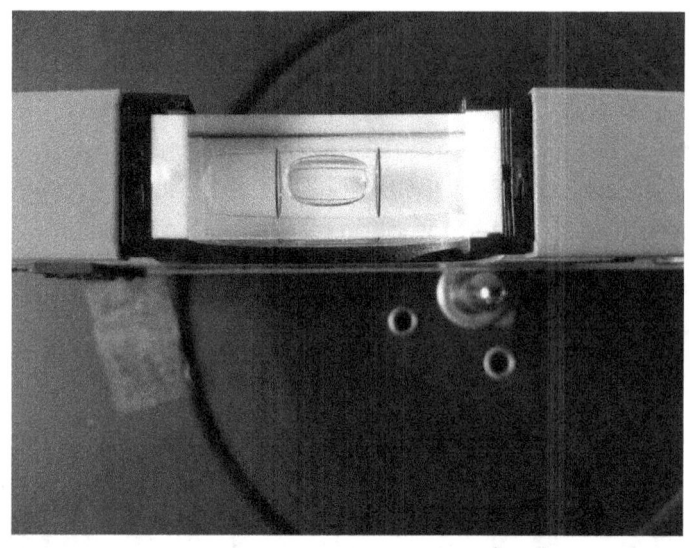
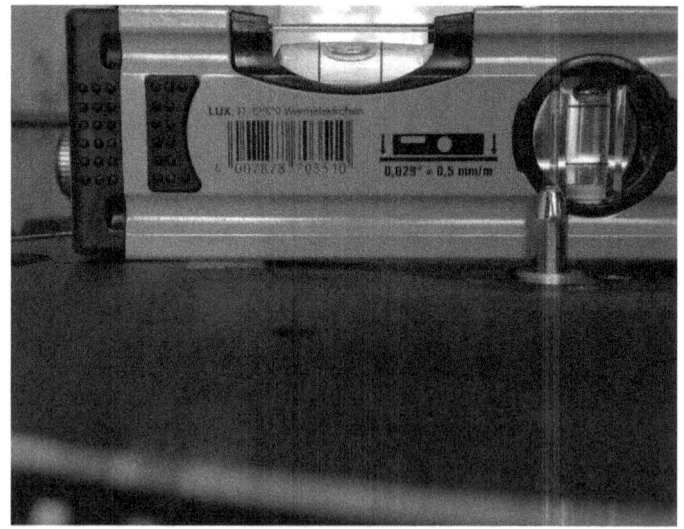

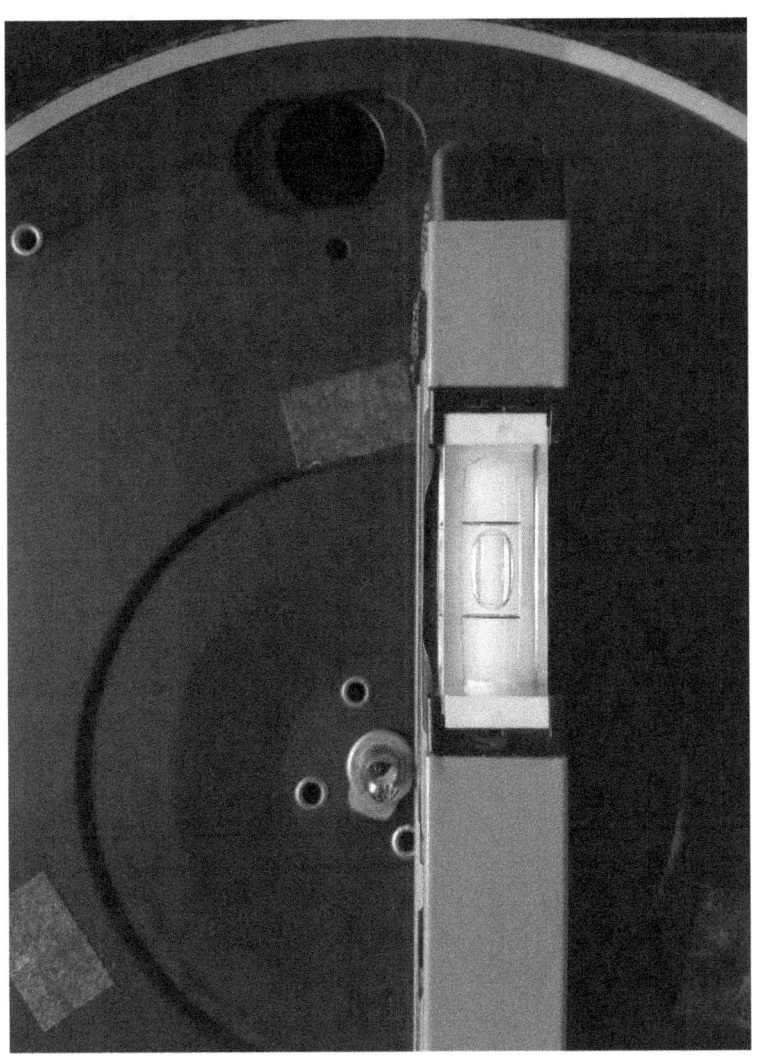

2. Cartridge alignment

To check and tune up the correct cartridge alignment (in order to be sure that the sylus is positioned properly relative to the grooves on the record), a cartridge alignment protractor must be used.

In this way, it's possible to minimize the tonearm tracking error.

It's possible to buy different types online, or even download a free file and print it, to proceed with the alignment.

Cartridge alignment protractor pic1

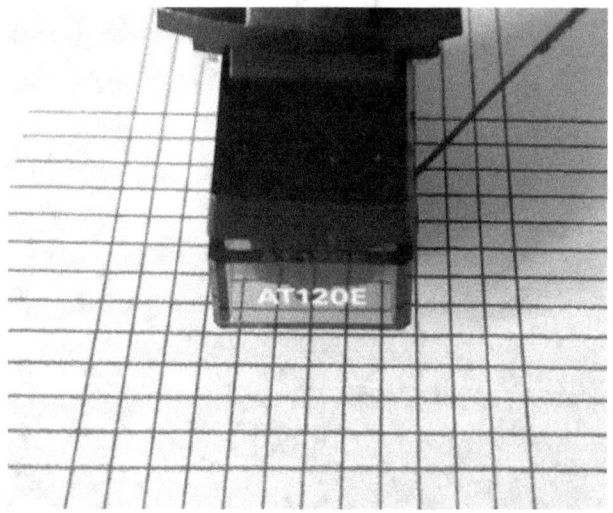

Cartridge alignment protractor pic2

Each protractor has two points, known as null points (points where there isn't tracking error) and at the end of the alignment tune up, the cartridge must be perfectly aligned in both points.

It's necessary to act on the overhang adjustment (the distance between the stylus and the center of rotation of the plate).

Overhang

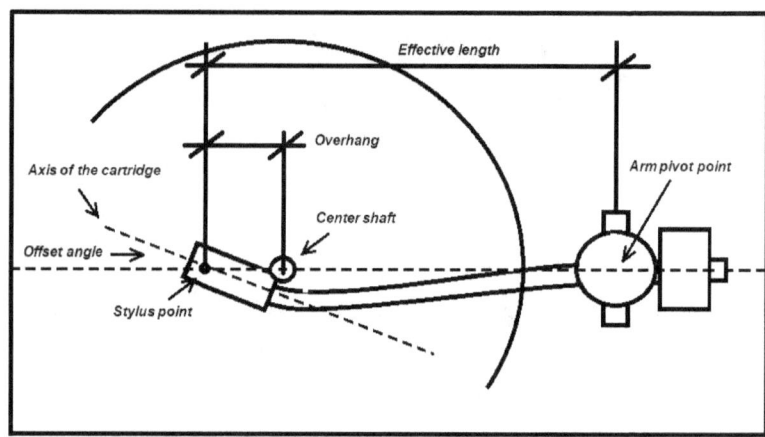

If the cartridge, once aligned to point 1 tends to point towards the outside of the record when moved to point 2, it must be moved back one the headshell, decreasing the overhang and it will be necessary to start again with alignment adjustments from point one.

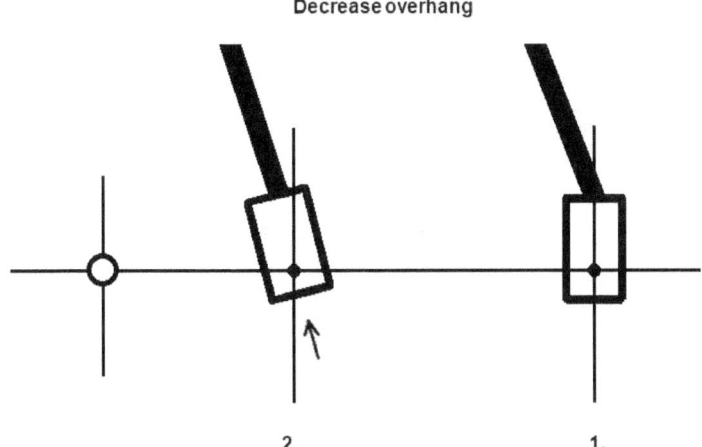

Decrease overhang

2. 1.

Conversely, if once aligned the cartridge with point 1 tends to point towards the inside of the record if moved to point 2, it must be moved forward on the headshell.

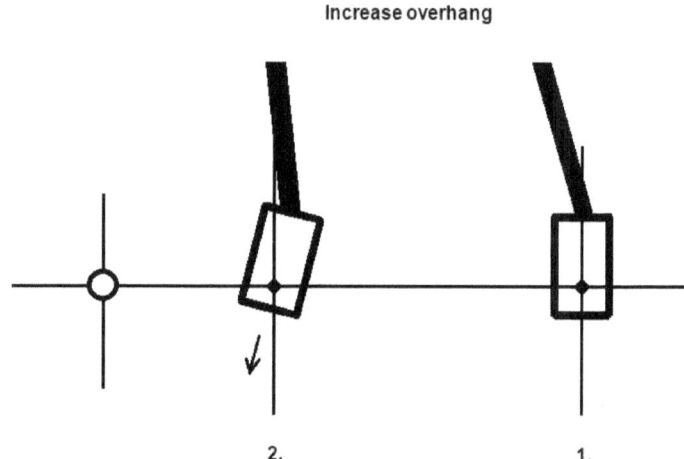

You'll have to proceed with the tune up, until the cartridge is perfectly aligned both in first and in the second point.
In this way, you'll have a correctly calibrated turntable (according to the studies of Bärwald and Löfgren).

For example, if you listen to a turntable with your headphones and the cartridge is perfectly aligned only with the point 1, past listening of hald side of the record, you will hear the right or the left channel in your headphones defecting (depending on the wrong setting) as if the headphones had the damaged cable, or a drop in the volume of the left / right channel without you having changed the balance between the channels via the amplifier / preamplifier knob.

If, on the other hand, the cartridge is perfectly aligned with point 2, but not with point 1, you will hear correctly the songs engraved on the first half of the record side you are listening to and the final part with the defects listed above.

In this case, it will be necessary to double check that the cartridge is perfectly aligned both in point 1 and point 2.

The elliptical styluses are most affected by an incorrect alignment; so the alignment with this kind of stylus must be perfect in order to enjoy the best performances.

Due to their shape, the conical stylus have a greater tolerance respect to an inaccurate alignment.

Once you have finished leveling checks with the bubble level, and having aligned the cartridge with the protractor, you can proceed with the next step.

3. *Tonearm balancing*

After fixing the cartridge to the headshell, it's time to balance the tonearm.

Block the movement of the tonearm with an object; (in the photo below I used a battery).

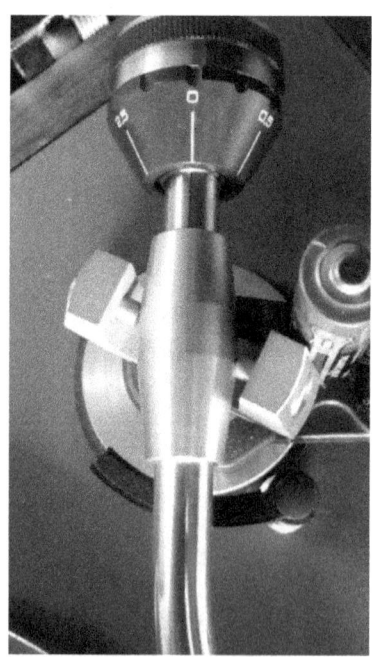

Lower the arm lifter, remove tonearm from arm rest and rotate the main weight to move it forward or rearward and obtain balance as in the picture below:

Perfectly balanced tonearm

After balancing the tonearm, return it to the arm rest and raise the arm lifter.

Turn the tracking force ring and set the reference line to zero, as displayed below:

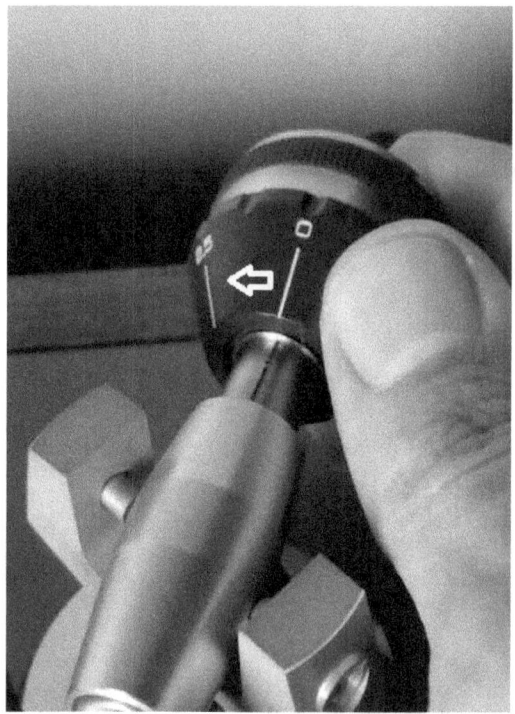

Next, turn the rear part of the main weight in the arrow indicated direction, until you reach the recommended tracking force for your cartridge.

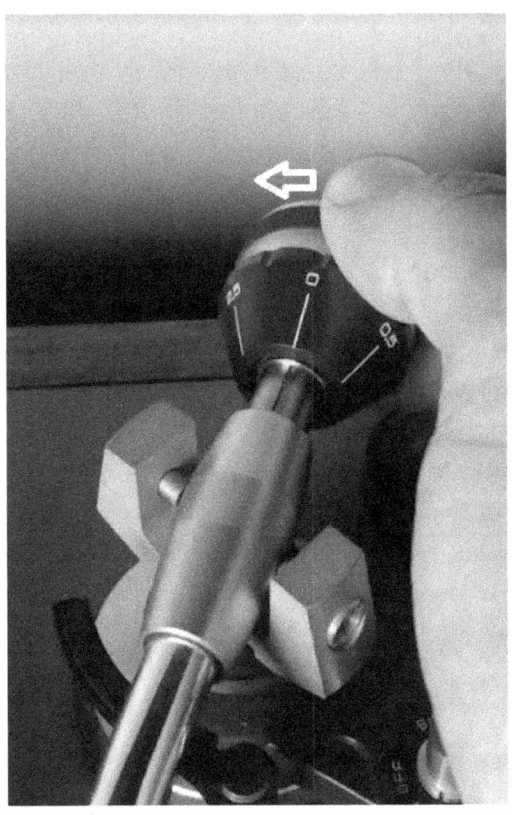

4. Antiskating adjustment

Antiskating is a mechanism that allows you to apply a very small force, to be adjusted according to the weight applied to the cartridge, for the correct tracing of the record.

The antiskating counteracts the force that tends to pull the stylus towards the center of the disc; it ensures uniform pressure of the cartridge on the right and on the left in the groove of the record (tracking force).

In this way, you prevent the stylus from wearing excessively only on one side and the greater consumption of a furrow wall.

To tune up the antiskating, it's necessary to set the relative knob, on the same value as the tracking weight provided by the cartridge manufacturer.

In some turntables, it's possible to make a very precise adjustment, since in addition to the weight applied to the cartridge, it's possible to select the type of stylus used (elliptical or conical).

In the photo below there are two graduated scales to regulate antiskating; the top one (in use) for the elliptical stylus and the bottom one for the conical stylus.

Usually the turntables have 2 speeds (33-1/3 e 45 rpm) and when adjusting the antiskating, it is always good to set the highest speed (45 rpm) since the highest speed, can put antiskating more in crisis.

To tune up antiskating, there are records on the market with a smooth track, and when the stylus rests on this track, the tonearm must not move towards the inside of the record, nor towards the outside; must remain still.

Record with a smooth track, to tune up antiskating

If you can't find a record with a smooth track, you can also use a damaged cd, with the label side resting on the plate as shown in the photo below.

The result will be the same!

IMPORTANT: if you use the cd, be careful to immediately catch the tonearm before it slides off the edge of the cd or towards the center shaft, in order to avoid to damage the cantilever (or put K.O. an expensive MC cartridge!). Only when the tonearm remains motionless, will be certain that only the effect of the groove spiral will move the cartridge and the tonearm from the beginning to the end of the record.

Another kind of record, on the other hand, has 3 tracks available in 3 different points of the record and by using a modulated signal, it allows the user to tune up the antiskating by ear.

In the tangential type tonearms, there is no phenomenon of pushing towards the inside of the record; for this reason, in this kind of tonearms there isn't antiskating adjustment knob.

Fine speed controls

In most existing turntables, there are separate knobs for fine speed adjustment.
With the platter rotating at the selected speed (33-1/3 or 45 rpm), observe the stroboscope.
Adjust the knob corresponding to the selected speed until the stroboscope pattern appears to be stopped.

Stroboscope

If you use accessories such as an antistatic record cleaning arm (to eliminate dust before it reaches the cartridge while reading the disc), or a clamp (to make the record perfectly flat while being played and absorb vibrations), proceed with the fine adjustment of the speed after placing these accessories on the record.

Antistatic record cleaning arm

Clamp

Conclusions

Always use a stylus suitable for your turntable.
Same thing for the cartridge and the tonearm on which it's mounted; their weight, their adaptability and adjustments play a leading role, to the point that they determine the rubbing coefficient on the walls of the furrow.
On average, a stylus needs a 30/40 hour break-in period to ensure that it can provide the best performance.
Pay attention to the wear of the stylus, in fact a damaged stylus can damage, in a few evenings, an entire collection of records!
A stylus of this kind does not have as a sole consequence the decrease in listening quality, but manages to affect and mark the precious sulcus material.
If possible, always keep a spare stylus and do not hesitate to replace it, or at least to have it checked if you notice a drop in sound performance.
It's estimated that a normal stylus should be replaced after reading hundred 33 rpm records and a diamond stylus after a thousand 33 rpm records.

About the load capacity (pF), impedance (kOhm) and output capacity (mV)

In some preamps/ integrated amplifiers, there is a switch that allows you to select the most suitable load for the cartridge in use.

The switch on the left, allows you to adjust the impedance of the cartridge (expressed in kOhm), the one on the right the load capacity (expressed in picofarad).

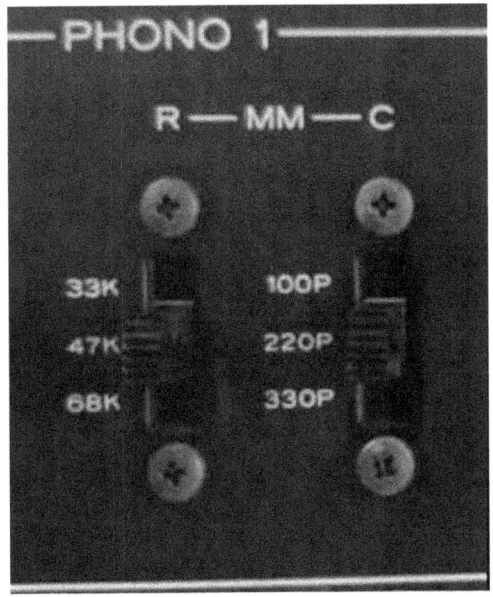

The instructions / characteristics of the cartridge, indicate the load capacity recommended by the manufacturer.

In the load capacity of the cartridge, you must include also the capacity of the tonearm's cables and the audio output cables of the turntable; since wiring harnesses are different from turntable to turntable, the load capacity should be calculated on each turntable.

However, we tend to assign an approximate value of 150pF to the turntable.

Let's assume that the cartridge has a recommended load from 150pF to 200pF.

It will be necessary to consider 100pF on the preamplifier / amplifier and 150pF for the turntable (a more or less standard size), for a total of 250pF.

It's not important that the load matches perfectly, it's usually correct if the value is within the +/- 100pF recommended by the manufacturer.

Let's assume another example; cartridge with recommended load of 400pF.

We will consider 220pF on the preamp / integrated amp and considering about an approximate 150pF for the turntable, we'll get 370pF.

Some of the best preamps / integrated amps, take into consideration the collection of cartridges of the enthusiasts who will buy them and through the available settings, allow an almost perfect matching with any type of cartridge.

In the picture below, the "phono1" section offers four output adjustments positions; 1mV, 2.5mV or higher (MM cartridges), 250µV and 100µV (MC cartridges).

Therefore it can be used with low or high output MM cartridges and with low or high output MC cartridges, eliminating any possible incompatibility between the output of the cartridge and the input sensitivity of the preamplifier ensuring an excellent signal to noise ratio.

The "phono2" section, is designed to manage only MM cartridges with an output capacity of 2.5mV or higher.

The input impedance of the MM cartridge is 47 kOhm, and 100 / 220 Ohm for MC cartridges, depending on the selected MC input.

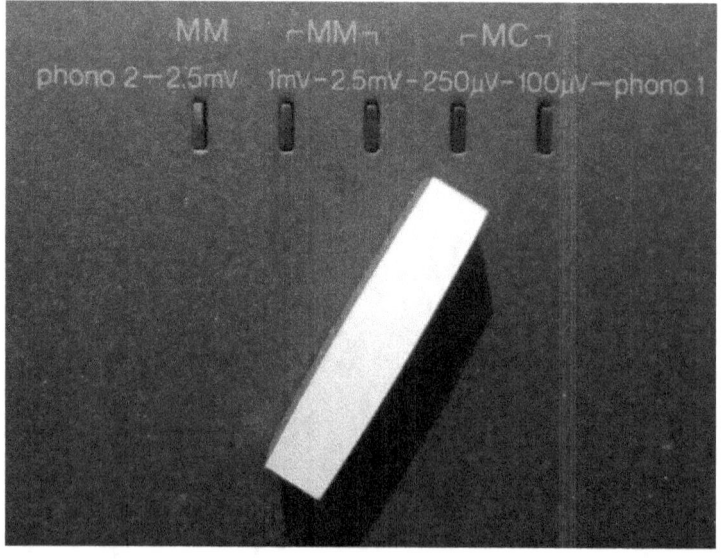

The network audio player

Although it's not a "real hi-fi component", like tuners have been in the past, many manufacturers of audio components have decided to introduce them among their products, since many people enjoys listening radio programs or their playlists.

They are the evolution of the old tuners (wich in the past allowed the listening of radio programs with a certain audio quality) but with a "virtually infinite" library of music files.

Through an internet connection, they allow both listening to internet radio stations and your favourite music apps streaming; there is also the possibility to connect and stream directly from your computer, external HDD, or to your portable mp3 player (phone, tablet, etc.), allowing you to listen to high resolution audio files.

These kind of components are in general more or less equivalent; in this case the audio quality will depend on the quality of the internet connection, internet radio broadcasting, and also on the resolution of the files you're going to play.

Some network audio players are connected to the internet via ethernet cable, others via wi-fi connection.

A network audio player connected via ethernet cable

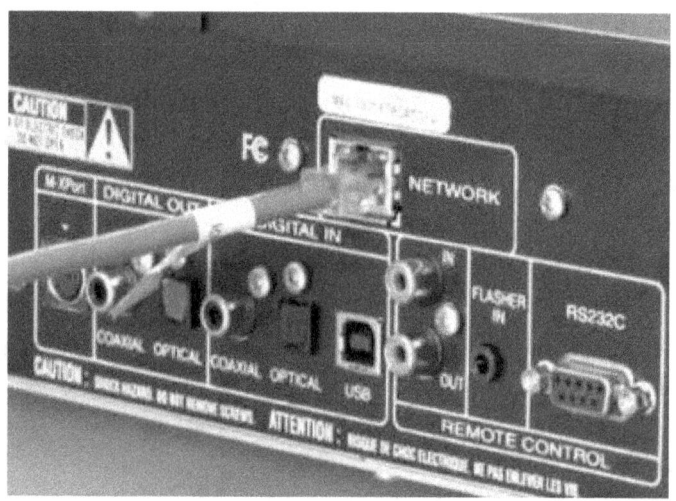

Chapter III
The Amplifier

The amplifier is a fundamental component of the audio chain.

In fact, it must receive the audio signal from the selected source (for example a turntable, CD player, tuner, etc.) and carry the audio signal without adding noise along the path to the speakers.

It can be integrated, when the preamplifier and power amplifier section are integrated within a single frame.

Otherwise, the solution in two frames (preamplifier and stereo power amplifier) increases quality and performances in terms of sound quality and power, or even in three frames (preamplifier and 2 mono power amplifiers, one for each channel).

All this, has the pros and cons; the integrated amplifier certainly has a lower cost, and needs less space to be placed although it has a lower quality than a two-frame solution (pre + power amp).

From a quality point of view, the solution in two/three frames is certainly higher, as is the cost (to which is added also that of an additional cable to hook up the

preamp with the power amp, which must be of adequate quality to the electronics to which it will be connected);
it will certainly also need more space to be placed.
The chances of a fault also increase, as there are a greater number of electrical/electronic components for signal processing.
If you are going to purchase a pre + power amp used, always make sure that the preamplifier has the power amplifier provided by the manufacturer for that model and vice versa; this is very important to maintain the balance foreseen by the manufacturer during the engineering and subsequent development, as well as allowing the electronics in two / three frames to best express their potential.
If you initially find and decide to buy only one of the two / three frames, please be patient and take your time to retrieve the other missing component (pre or power-amp), in order to allow the perfect electronics matching and enjoy the sound produced by that combination, to understand the kind of result that engineers and the manufacturer wanted when these frames were produced.

As a general rule, there is no hi-fi brand superior to the others. During "hi-fi history", all manufacturers have produced astonishing electronics, as well as rubbish electronics, so if you aren't an expert, you need to document yourself and whenever possible, carry out listening tests.

In fact, even specialized amplifier brands have had difficult moments (economic, or change management), producing poor products or in any case not up to the brand reputation and its history.

Another important factor for choosing the amplifier, is the size of the listening room where it will be positioned.

If your budget is high and the room is large, you can buy amps with hundreds of RMS watts of power.

Otherwise, it makes no sense to buy a power amp capable of soundproofing a theater or an auditorium if you have a small room dedicated to listening, because you will never be able to fully enjoy your jewel.

If, on the other hand, you have a very large listening room, a sound system with scarce power capacity could distort the sounds.

Here are some positioning tips:

Remember not to obstruct the top, side and rear grids of your amplifier by placing books, other audio equipment, etc. on top of the unit in order to prevent it from overheating.

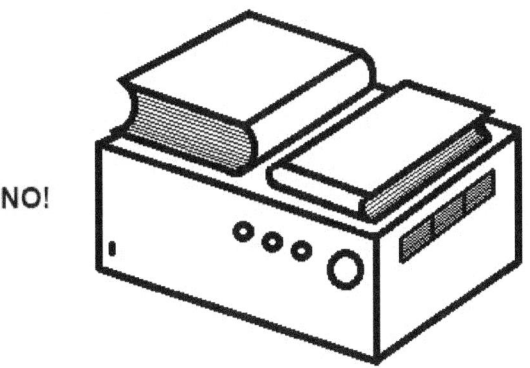

NO!

Leave at least 4" (10 cm) between the unit and a wall, cabinet or other equipment.
Do not expose to direct sunlight.
Place it where humidity is low, on a flat and free from vibrations surface; away from stoves, radiators and other sources of heat.

IMPORTANT: if you have purchased a very powerful amplifier, make sure that the volume knob is <u>always turned to minimum before switching on the unit</u> (especially on vintage equipment).
Otherwise the unit and your speakers may be damaged.
After selecting on the preamplifier / integrated amplifier the source you want to listen to (CD player, turntable or other), gradually increase the volume until you hear the sound.

Chapter IV
About amplification and "Clipping"

Is it better when the amplifier is more powerful than the loudspeakers, or when the loudspeakers are more powerful than the amplifier?

Many people have asked themselves this question (myself included); let's better understand the topic.

It's important that the signal reaches the speaker without distortion.

In fact, there is the risk of damaging the speakers both by driving them with excessive power, and by driving them with not enough power.

Let's take for example a couple of 40 continuous watts rms loudspeakers (Root Mean Squared = RMS watts) each (with a maximum peak of 50 watts rms) and let's exam them with two different amplifiers.

The first, an integrated 26 watts rms per channel, and the second a power amplifier (combined with a pre-amp) of 120 watts rms per channel.

The loudspeakers will play with both of the amplifier mentioned earlier, however they'll reach their maximum potential (40 continuous watts rms) only with the 120 watts rms per channel amplifier.

In particular, with the 26 watts rms amplifier, you must be careful not to "clipping it", situation that occurs when an amplifier sends a distorted signal to the loudspeakers.

If the loudspeakers receive a distorted signal, the first component to be damaged will surely be the tweeter (the speaker dedicated to the reproduction of high frequencies) and then the remaining others.

How to recognize then, which is the maximum volume to "push" our 26 watts rms amplifier without clipping it?

In fact, we usually don't use the amplifier inside a laboratory, connected to test equipment to be able to evaluate its performance in real time.

Here is the answer; often (not to say always) amplifier manufacturers set the clipping threshold when the volume knob reaches the one o'clock position.

Overcoming this position with the volume knob, you are warned that you're sending a distorted signal to the loudspeakers. This rule is valid for any existing amplifier, whatever power it has.

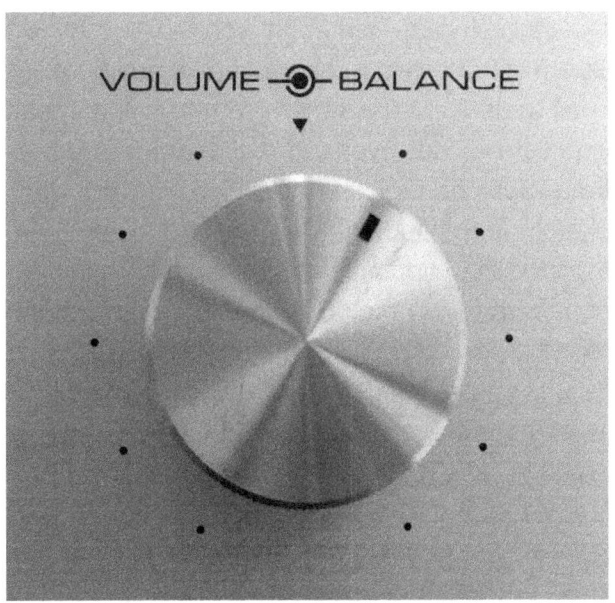

Imagine a party scenario or while listening to the album of your favorite singer (which you have the pleasure of listening to at high volumes), where the volume knob gets out of hand and significantly exceeds the position shown in the photo.

Thinking about putting out sound volumes that your small amplifier cannot handle, although you have high efficiency speakers (I will elaborate in depth the concept of efficiency later on), you will certainly damage your beloved speakers!

Feeling safe to not overload the speakers with a simple 26 watts per channel rms amplifier, you will destroy the speakers due to clipping.

By respecting the clipping threshold, you'll avoid returning the speakers to the factory (within the warranty period, in case of brand new loudspeakers); this is what really happened in the 80s to an Italian loudspeakers manufacturer who saw many speakers (tested by the factory with amplifiers with higher power than requested) coming back to the factory, without understanding why.

In reality were the customers themselves to damage

their loudspeakers due to their amplifiers clipping.

Regarding the second amplifier (the one with 120 watts rms per channel), once the loudspeakers are connected, you'll hear them playing really loud and with a power reserve that can be perceived already at a very low volume. The loudspeakers themselves, which seemed sluggish, will sound completely different, providing a greater impact through the music scene especially an orchestra in "full flight", rather than in rock concerts; they will transform that sound that with the previous amplifier remembered the braying of a donkey, in the song of a siren enchanting sailors.

Obviously you'll not have to exaggerate with the volume even with this amplifier, because of the more than double power supplied against that supported by the loudspeakers.

However you'll be able to enjoy your speakers to the fullest by using all the 40 watts rms even for long listening sessions without compromising neither the amplifier nor the components of the loudspeakers themselves.

Answering the initial question, it's definitely better to have a more powerful amplifier than the loudspeakers.

The ideal would be to have an amplifier with double power capacity than the loudspeakers, or even slightly more than double.

For example, 2 loudspeakers of 70 watts rms each hooked up to an amplifier with 140/150 watts rms per channel.

In order to reach the "middle zone" and/or the "clipping zone" of the volume knob, without fear of damaging both the speakers and the amplifier, being able to enjoy a good listening volume even for many hours in a row (if required).

If you really can't remember the limit threshold of the volume knob, you can always apply an adhesive "marker".

Another thing to check, when we are deciding which amplifier to buy is the power consumption.

In fact, the value of the current absorbed by the amplifier can give us a realistic indication of the real potential of the amplifier itself, since always an amplifier can't put out more power than it consume.

Just as supercars consume a lot of petrol to get high performances, amplifier also need a lot of current to "drive" any kind of loudspeaker both at low volume and at high volume without difficulty.

Amplifier classes in a nutshell

The way an amplifier combines power and signal defines its class. Let's make an overview about the most common circuit tipology: Class A, Class AB and Class D.

If I have a 100 watts **Class A** amplifier, when I plug that amplifier in and I have no musical signal at all, it's consuming more than 100 watts (as example drawing 140 watts) since <u>amplifiers always put out less power than they consume</u>.

It has a constant Bias (Bias current refers to the current is using the amplifier when is not receiving any musical input signal), that means that the amplifier with no musical signal is constantly on and drawing current producing heat.

When I put a musical signal into the amplifier I will still have only 100 watts going through the amplifier but instead of the wattage be converted to heat, it now goes to the loudspeakers and is converted to sound.

So at its highest musical signal level all of the energy is going to the loudspeaker, with a lowest musical signal level the amplifier is turning all the energy into heat but it's always consuming that amount of watts.

The reason we want to do that is to keep the transistor on all the time, in order to don't go through this on/off phase that a number of amplifiers do.

Class A amplifier <u>has the highest sound fidelity</u> with lower distortion, but is the least efficient since runs with constant bias; in other words, it always runs full power both with music signal input or not. When there isn't any music signal, power turns into heat, vice versa when there is a music signal input the power goes to the loudspeakers.

Now let's go on to **Class AB**.

Class AB is the same thing I described above but instead of drawing 100 watts and converting that to heat (also when the amplifier doesn't receives any music input signal) as mentioned in the earlier example, now we're gonna draw a fewer watts constantly.

So as example there is a constant Bias of 35 watts on current.

When I plug this amplifier in, and I don't put any musical signal into it, it's going to draw a constant 35 watts and this 35 watts will be transformed into heat; as the musical signal starts increasing, let's say up to 100 watts, the energy will go all to the loudspeaker.

Then, when the music signal stops, the energy stops going to the loudspeaker and the amplifier comes back to drawing 35 watts, and the Bias constant current is transformed into heat.

Summarizing, a 100 watts Class AB amplifier will work in Class A up to a certain amount of Bias current (in this case 35 watts). When will exceed 35 watts (from 36 up to 100 watts) the amplifier will work in Class B.

Class AB amplifiers, combines the sound quality of Class A amplifiers until a certain amount of power and offers a better power efficiency than a Class A when there isn't any input music signal (since it draws only 35 watts against 100 Class A watts!).

So now let's move on to **Class D**.

Class D uses "pulse width modulation" or PWM, and that's where we have a stream of pulses that are coming out at a regular frequency (often over 100K Hz), that means once every one hundred thousandths of a second there's going to be a pulse and depending on how loud the signal is, that pulse will either get wider as the signal gets louder, vice versa as the signal gets quieter that pulse gets lower.

Since the transistors that make the Class D output stage are either on or off, they are running at their maximum efficiency so very little heat is generated.
Class D amps have very good efficiency but aren't "Hi-fidelity" components.

Chapter IV
The Loudspeakers

Here we are! At the loudspeakers, the true "voice" of the stereo system.

It often happens to read reviews, in newspapers or on forums / social networks that praise loudspeakers with unusual qualities, while they denigrate other loudspeakers that perhaps don't look good but sound really good.

Technology made giant leaps forward, and it's possible to find many models on the market with the most varied characteristics.

As a general rule, I inform you that the american loudspeakers are among the best in the world; so if you are planning to buy american loudspeakers, you're on the right path.

One of the fundamental parameters to get satisfaction from your loudspeakers, is to establish the kind of music they will have to play.
In fact, it's rare for a loudspeaker to be suitable for playing any kind of music.

The first thing to check on a loudspeaker is the **frequency response**.
Let's assume, for example, that we have seen a pair of loudspeakers on a catalog, which visually inspires us to purchase.
We take a look at the technical specifications and the manufacturer declares a frequency response from 50 Hz to 17.000 Hz.

The frequency response indicates that these loudspeakers are incomplete, since most of the music we listen to, goes far below 50 Hz and goes beyond 17.000 Hz.

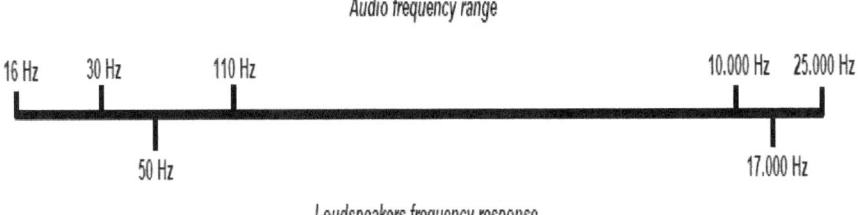

Loudspeakers frequency response

For example, if you listen to jazz, maybe you're missing out a bit, because some songs are focused on mids and highs, while others also go towards the low range; it depends on the jazz you choose.

If you listen to rap, you're losing a lot, if you listen to hip hop, you're losing a lot, if you listen to blues, you're losing a lot, if you listen to country music, you are losing a lot, if you listen to rock, you are losing a lot, if you listen to metal or hard rock, you are definitely losing a lot, if you listen to classical music, you are very far away, in fact a lot of classical music goes beyond 25.000 Hz and goes down to the low range up to 16 Hz.

With loudspeakers like these, you're listening to music for a half...!

You're missing out on all the pleasure a piece of music can give you with its tonal details.

In practice you aren't listening to the song in the way the producer and his artists intended to be heard later, as when it was recorded in studio.

These speakers are unable to reproduce the music scene in the same way it was recorded in studio; therefore they are incomplete speakers because they don't reproduce the sound correctly.

Probably, they are able to do their job well only in a restricted musical area, such as female voices (a cappella) and wind instruments.

It's a bit like if you inserted the CD of your favourite rock band, and you eliminated the drummer from the band because the speakers aren't able to correctly reproduce part of the low frequencies produced by percussions and the high frequency range of the cymbals, which emit a frequency above 17.000 Hz.

The first note of the organ pedalboard corresponds to 16.35 Hz.

In England organs have no pedalboard (with rare exceptions) and therefore the British aren't interested in playing the first octave (not present in their music).

An important English loudspeaker manufacturer, over the years, had to change the frequency response curve in the low range of its flagship pair of speakers because it starting from 60 Hz, cutting out all the music with low range frequencies between 16 Hz and 60 Hz.

The buyers (from other parts of the world) of these luxurious english loudspeakers, had noticed this shortcoming by signaling it to the manufacturer that, as just mentioned, had to run for cover improving the frequency response.

Below is a list of some of the low range frequencies:

E = 80 Hz lower string of the guitar
E = 40 Hz lower string of the double bass
C = 32 Hz first note of the organ keyboard
A = 27.5 Hz first note of the piano keyboard
C = 16.35 Hz first note of the organ pedalboard

To better understand the "frequency cut off" speech, you can listen to some of your favourite's rock band songs through the speaker of your smartphone or through the small speakers of your laptop which usually have a frequency response ranging from 100 Hz to 12.000 Hz approx.

Since the small speakers will not be able to reproduce the low range and also the higher frequencies, the piece of music you're going to play will loose a large part of the musical information conained in the song... therefore, it will be like listening to it for a half.

To consider a pair of speakers as valid, keep in mind a frequency response that is between 30 Hz and 25.000 Hz.

Although there is always a cut of the low frequencies between 16 Hz and 30 Hz, you will be able to reproduce at best almost all the existing music and without major losses in sound fidelity.

The second thing to check regarding the choice of a speaker is **sensitivity**.

The sensitivity of a loudspeaker is the sound pressure level detected to one meter from the speaker when a 2.83Vrms signal is applied to its input.

The value of sensitivity makes us understand how much more/less efficient a speaker is; that is, how much electric power a loudspeaker needs to reach a certain sound pressure.

More simply, efficiency is the ratio between the input power and the output power.

If our amplifier pumps 30 watts rms into a loudspeaker to obtain 5 watts rms of sound power output, surely the loudspeaker isn't very efficient (this without considering the frequency response and sound irradiation capacity, since these two factors also impact on the efficiency of the loudspeaker).

A low efficiency loudspeaker has the defect of wasting part of the energy that receives from the amplifier.

The failure to transform the received electrical signal into sound that is drained/transformed by the loudspeakers in heat (which can also damage them).

On the other hand, very efficient loudspeakers hooked up to a 60 watts rms amplifier can sound louder than another amplifier with 120 watts rms power output hooked up to poorly efficient loudspeakers, because they need little current to put out a great sound.

If you'll choose an amplifier with a few watts, you'll have to orient yourself towards loudspeakers with good efficiency; if you'll choose an amplifier with hundreds of watts, loudspeakers with poor efficiency will not be a problem.

Below a classification of loudspeakers based on sensitivity:

- less than *82 dB*; very low, only for small rooms or close listening
- *82-84 dB*; low
- *85-90 dB*; average
- *91-95 dB*; medium high
- *96-99 dB*; high
- from *100 dB* and over; very high

A good sensitivity value for a pair of loudspeakers is between 89 dB and 92 dB; you can certainly interface your loudspeakers without problems with the most of amplifiers currently on the market, both vintage ones.

A lower sensitivity value, compared to the range mentioned above, increases the criticality of coupling with the amplifier, reducing the variety of choice.

Third thing to check is **impedance**.
Indicates the current load on which a specific pair of loudspeakers works and indicates to the amplifier on which load it'll have to supply current.
The most common impedances are 8 Ohm and 4 Ohm (less common 6 Ohm).
The more the impedance decreases, the more the amplifier's driving criticality increases.
Hooking up loudspeakers with 4 Ohm impedance, you can perceive that your amplifier producing more heat (since it's engaged on the most difficult load) and if the amplifier doesn't feel particularly comfortable during a demanding musical performance, you may hear a volume drop or a sound interruption since it's gone into protection.
This means that your loudspeakers need a more powerful amplifier to be better driven.
Today all amplifiers can handle easily 4 Ohm loudspeakers, if instead you have in mind to buy a vintage amplifier you'll need to collect some informations, because in the 70s and 80s not all the amplifier on the market were able to

drive correctly loudspeakers with an impedance of 4 Ohm. To avoid headaches, and to avoid scrambling (or worse, damaging) your amplifier, I recommend you to choose speakers with an impedance of 8 Ohms.

Summarizing what has been said so far, a loudspeaker is critical to combine with the amplifier when it has a low sensitivity and low impedance. Conversely, it's easy to combine, when it has a medium-high sensitivity and a high impedance.

Fourth thing to check, is the **power**.
Important premise; buying a 250 watts rms loudspeaker will not make the sound system playing louder if your amplifier puts out only 100 watts rms.
Not only could risk damaging the loudspeakers, clipping the amplifier, but you would not even be able to enjoy the full potential of the loudspeakers, because driven by an amplifier not too powerful.
As suggested earlier (on page 76), recommend that the amplifier have at least twice the power of the speakers, so that you can enjoy the maximum sonic performances (obviously keeping attention to don't reach the "clipping area" with the volume knob to avoid loudspeakers damage due to an excessive heat accumulation).

Regarding the power, it's important to know the size of the room; How big of a space do you have? Will you use the loudspeakers in a very large living room, or in a small studio?
Once measured the size of the room, you will decide

how much power apply.

For a small room, a pair of loudspeakers up to 50 watts rms will be ok.
For a medium sized room, a pair of speakers with power between 160 and 200 watts rms will be fine.
For a large (or very large) room, a pair of loudspeakers with power between 250 and 400 watts rms and over will be perfect!

Other fundamental parameters in a loudspeaker are its **size** and **kind of loudspeaker, kind of project** and **number of ways**.

Some time ago, I read a gentleman's comment on a social network, which boasted of the "excellent sound" of its speakers positioned on stands.

A couple of minutes later, the following comment from another gentlemen arrived: "what sound do you want them to make, if they are sized like a shoe box?!?!"

Severe but fair comment.

In fact, although a small loudspeaker can mount excellent quality speakers, due to its small size, it will have difficulty reproducing an orchestra in full flight, rather than the reconstruction of a large pipe organ inside a cathedral (just to mention some examples).

So a bigger case will surely have superior abilities to generate a "great sound", reconstructing at best the sound scene, especially in the low frequency range.

After this nice anecdote, we can get to the heart of the topic, dividing the high-fidelity loudspeakers basically into two categories: **floor** and **bookshelf**. (There would also be 2.1 loudspeakers, that is, two satellites together with a subwoofer, which however I consider more suitable for listening to the audio of a laptop without big claims, due to the physical limitations of the satellites, poorly suited to reproduce a high fidelity sound and therefore not recommended).

Usually the floor loudspeakers are of a fairly important size and weight and often guarantee the best performances in general, especially regarding efficiency, power and frequency response; this kind of loudspeaker is certainly more demanding both in terms of costs and in terms of the space occupied inside the room dedicated to listening.

The bookshelf loudspeakers are less bulky and less heavy; this lesser "phisicality" certainly determines a lesser impact in the reconstruction of the sound scene, and also a lower performance level than the floor ones.

It follows, therefore, that even by applying a high power to library speakers (also of excellent quality), these will never be able to approach the level of a good pair of loudspeakers with important dimensions.

If your listening room is not very large, and can only accomodate bookshelf loudspeakers, pay close attention to the quality of the speakers they mount, in order to guarantee a sound intensity suitable for the kind of music you listen to more often and check especially that the woofer is large enough to reproduce the lower frequencies.

IMPORTANT: You can buy the best loudspeakers in the world, but if you don't have the patience to perform listening tests with different loudspeakers positions, they will always sound bad.

The floor loudspeakers, must be positioned far from the back wall, from the side walls (or you'll obtain a bad reinforcement of the low frequencies, in particular with loudspeakers with the "reflex tuned port" in the rear part of the cabinet), and as far as possible from the audio source (especially if you use a turntable) to avoid the phenomenon of the acoustic feedback, that is when the vibrations and sounds of the environment are perceived by the source and added to the output sound of the source, altering the reproduction quality of the support in use (vinyl record/cd/etc.).

Regarding the type of cabinets, here is a list of some examples:

- **Horn** loudspeakers; these are the most efficient speakers ever. The flared shape of the trumpet in front of the speaker generates the effect of a megaphone positioned in front of the loudspeaker. They were used with the first tube amps, which had only a few watts of power.
They are very bulky and their use is usually suitable in PA (pubblic address) broadcasting systems such as discos, stadiums, auditoriums and theatres where the primary factor must be the quantity of sound produced, rather than quality.

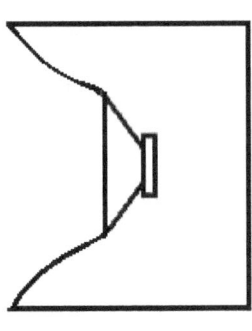
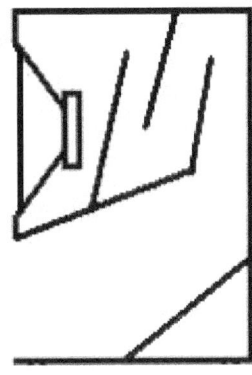

Horn loudspeaker   *Backloaded horn*

- **Infinite baffle** loudspeaker; it consists of a closed box, and since the internal air is completely isolated from the external one, there is no phase cancellation. They are probably the best fidelity loudspeakers, as regards the reconstruction of the sound scene in a domestic environment.
They aren't very efficient, and therefore they need to be driven by high power amplifiers.

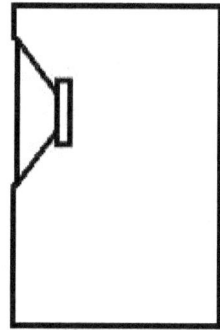

Infinite baffle

Another variant of the "infinite baffle" or "closed cabinet" loudspeaker is the use of a *passive radiator* (or A.B.R., auxiliary bass radiator) to increase the frequency range in the lower frequencies and also overall power of the loudspeaker.

The radiator consists of a sort of speaker, in which however, the electrical part that should make it move is missing; works in tandem with the woofer in the rear.

A function similar to the "tuned port" of the bass reflex, which releases the air produced by the rear of the woofer to increase the efficiency of the speaker but avoiding adding the distortion caused by the turbulence generated into the "tuned port".

This kind of solution, guarantees high-level sonic performances, however it's penalized by the high cost.

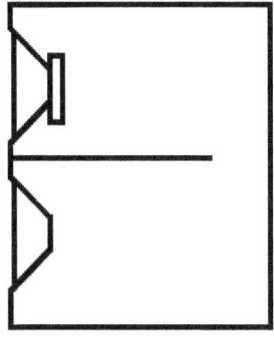

Passive Radiator

- **Finite baffle**; as the name indicates, the cabinet has an opening on the back.
It's used in instruments that have built-in amplifiers and speakers.
It's a little more efficient than the "infinite baffle" type, but poor in terms of low frequency response.

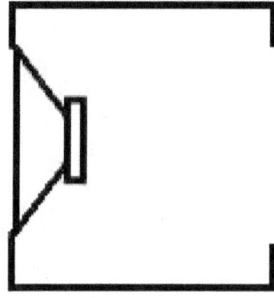

Finite baffle

- **Bass reflex** (or tuned port); the low frequencies coming from the back of the speaker, are put in phase through a "tuned port", and conveyed towards the front or the rear of the loudspeaker.

In this kind of loudspeaker, the size of the speaker is very important in order to reproduce the lower frequencies.

The efficiency is slightly higher than that of the loudspeaker in a closed cabinet.

Although this kind of loudspeaker sounds good, through the passage of air in the "tuned port", a turbulence is created which adds up to the sound coming out of the speaker; this slightly worsens the musical timbre of the music being played.

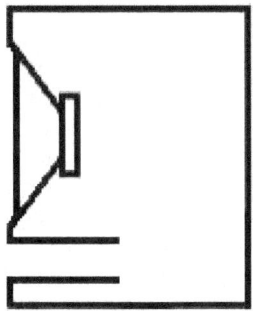 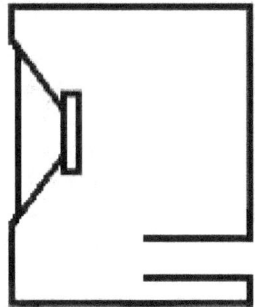

Bass Reflex "tuned port" – front / rear

Isobaric system

This kind of system, uses a second woofer mounted behind the one installed on the front panel and acoustically coupled to it via a sealed chamber.

The front and rear drivers are driven electrically in parallel and the rear driver, takes care of maintaining costant the pressure in the chamber between both drivers in order to allow the front operating driver to operate under a loading of constant pressure (external and internal to the loudspeaker).

Isobaric (derived from the therm Isobar "equal pressure"), is meant to explain that the pressure is maintained costant both in front and the rear of the external driver.

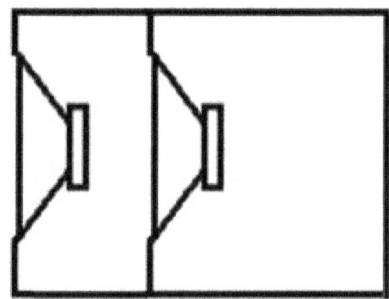

Isobaric loudspeaker enclosure

This solution is used to obtain two benefits when it's used a small cabinet.

The first one is a great control of the bass frequencies, since the front woofer is controlled by two coils.

The second one is that the internal woofer acts as if it were installed in a cabinet with twice the volume of the actual volume, increasing the extension of the low range frequencies.

Electrostatic loudspeakers

Bulky... expensive... difficult to drive due the lot of power requested... difficult positioning in the listening environment, as they require walls without furniture to get the best performances... no "dynamic punch" on bass frequencies and no impact... but they have also some flaws!

It works in a fundamentally different way than a conventional loudspeaker.
A thin electrically positively charged diaphgram flexes between two electrically negative conductive grids, and this produces the sound.

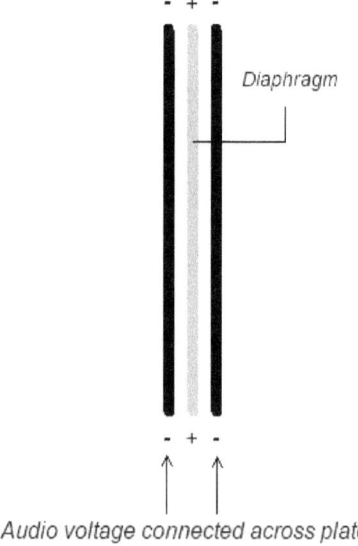

Audio voltage connected across plates

This kind of loudspeakers, can reveal music details unlistened before from very well recorded CDs, since they have great musical transparency, a really natural sound and a low distortion.

Are recommended for jazz quartets ensemble (or little jazz bands), very good also with classical music (except great orchestras).

Since they have a lack of loud bass, as mentioned earlier, aren't recommended for kinds of music such as disco, hip hop, rock/hard rock, and in general those kind of music that generates high sound pressures.

Wireless? No, thanks!

Being a book that deals with high fidelity, I wanted to dedicate a few lines to wireless speakers too.

Although in recent years the prices of these speakers have decreased, with an increase in the general quality, they still remain low quality speakers compared to a traditional loudspeakers as they are penalized by the transmission of the wireless audio signal.

Many people think that these wireless systems do not need wiring; this is an incorrect information.

In fact, although these loudspeakers do not require signal cables to be connected to the amplifier, they need to be connected to the electric network to be powered.

The money spent on the purchase of the rest of the high-level audio chain will be wasted if you insert wireless speakers at the end of the chain.

The main negative feature of this kind of loudspeakers are the interferences that afflict the wireless signal and that can manifest and annoy, while listening to a music song.

Full range, 2 ways, 3 ways or more?

Inside each loudspeaker, there is an electronic filter (crossover), which deals with dividing the frequencies of the sound spectrum. This filter, after cutting the audio spectrum in several parts, sends them to each dedicate speaker to reproduction.

Example of a 2-way loudspeaker:

    High frequencies → Tweeter
    Middle and Bass frequencies → Mid - Woofer

Example of a 3-way loudspeaker:

    High frequencies → Tweeter
    Middle frequencies → Midrange
    Bass frequencies → Woofer

Example of a 4-way loudspeaker:

    High frequencies → Tweeter
    Middle frequencies → Midrange
    Mid-Bass frequencies → Mid-Woofer
    Bass frequencies → Woofer

The list of examples could also continue as there are also 5-way speakers, or more.
However, in my opinion, it's better to move towards a good pair of 3-way loudspeakers, or at most 4-way ones. This for two reasons: the first is that everything that is not there, doesn't break (a greater number of ways, requires a greater number of speakers).
The second reason is the relationship between the exorbitant cost of more complex loudspeakers and the sound performance they achieve (equal or sometimes higher than excellent 3 or 4-way loudspeakers).
There are also full range loudspeakers, but since a single speaker is not able to reproduce the entire sound spectrum, it's an option to be discarded beacuse with a single speaker we aren't talking about high fidelity.

Here is an example of the frequencies on which the various speakers work, wich influence the cut of the crossover to which they are combined:

Tweeter   between 6.000 Hz and over 20.000 Hz

Midrange   between 350 Hz and 8.000 Hz

Woofer   between 30 Hz (or less) and 350 Hz

Regarding the tweeters, which are the smaller speakers, the dimensions aren't important.
The manufacturers tend to keep the midrange of small dimensions (10/12 cm) in order to allow the speaker to respond quickly to the most demanding musical passages.
Concerning the woofer, I suggest a size of at least 10" (25 cm), so that the membrane is able to move a sufficient mass of air and at the same time provide an extended response in the low range.
Loudspeakers with 12" (30/32 cm) woofers are also very good, since give back high sound pressure (also depending on the loudspeaker case capacity) and are able to go deeper in frequency (without losing articulation in fast music passages; this measure is an excellent compromise between the response speed of the membrane and the mass of air it can move).
The 15" (38/40 cm) woofers, they are able to return very high sound pressures, but with very high volumes they lose articulation, worsening the sound quality (especially in fast musical passages and in classical music; for this reason, they are more suitable for rock, dance or rap music).

Chapter V
Interconnect cables

Does it make sense to invest money purchasing quality signal cables?
Can an economic signal cable worsen the sound quality of the entire audio chain?
Can the section of the power cable between the amplifier and the speakers cause a decrease in signal strength?
The correct answer to all three questions is yes.

Signal cables, have always been the subject of discussion among hi-fi system enthusiasts.
Since they take care of transporting the signal along the entire audio chain, they can maintain the quality of the signal between the various components, or degrade the signal by adding interference due to insufficient shielding or decreasing the signal strenght in the passage between the amplifier and the speakers.
Let's consider the connection between the source and the integrated amplifier / preamplifier.

In the photo below, an example of the unbalanced signal cables that are usually found in the electronics equipment packaging.
Unbalanced cables just have two conductors, one running the audio signal and the other for the ground.

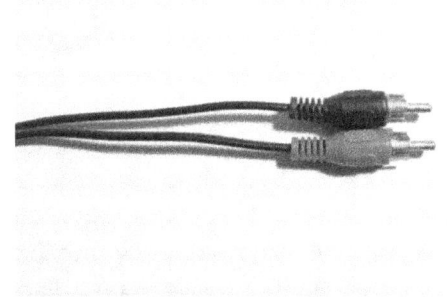

An economical **RCA** cable like the one in photo, connected between the output sockets of a source (ex. CD player) and the sockets of an amplifier / preamplifier, can pick up radio frequency interference which are added to the audio signal, by degrading the signal into more or less audible way.
As suggested on page 13, always replace the cables supplied by the manufacturer with quality cables.

In the photo below, you can see a shielded copper RCA cable without oxygen and gold-plated connectors.

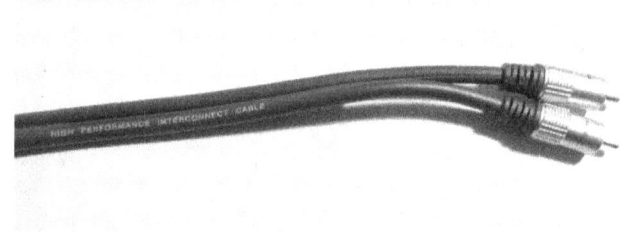

This cable can prevent radio frequency interference from entering the audio signal path, worsening its quality; even the gold-plated connectors help to keep the details of the signal to the amplifier.

IMPORTANT: If the cable is available in two different lenghts (for example 1 meter, or 2 meters), it is <u>always</u> preferable to choose the shortest one, so that the signal takes the shortest path between the two electronic equipments, greatly reducing the risk of collecting interferences along the route.

Below is the comparative photo between the two cables.

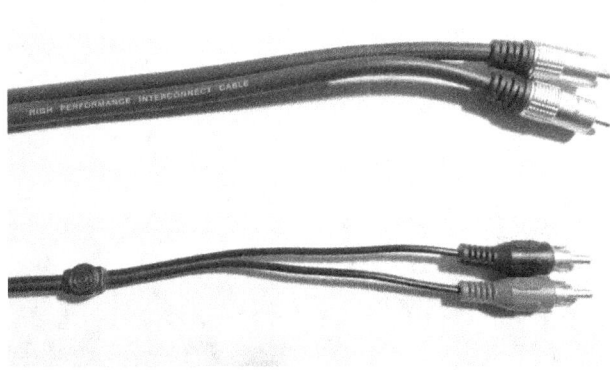

We immediately notice the difference in section between the two cables, due to the shielding of the high-performance cable.

The various constructive differences between the two cables do not improve or worsen the frequency response; no unbalanced cable is able to produce timbral variations on the audio signal during the passage.

However, the shielded cable keeps the signal quality high during the passage from the source to the amplifier, avoiding signal degradation.

In Hi-end devices (cd players and integrated amplifiers / preamps) and in the professional field, we find the **XLR** sockets (or even Cannon).

XLR sockets.

An XLR cable, in fact, allows you to connect devices at great distances (ex.over 10 meters) without loss of signal quality and is immune to the noise generated by radio frequencies, electronic equipment, etc.

Cable with female and male XLR connectors.

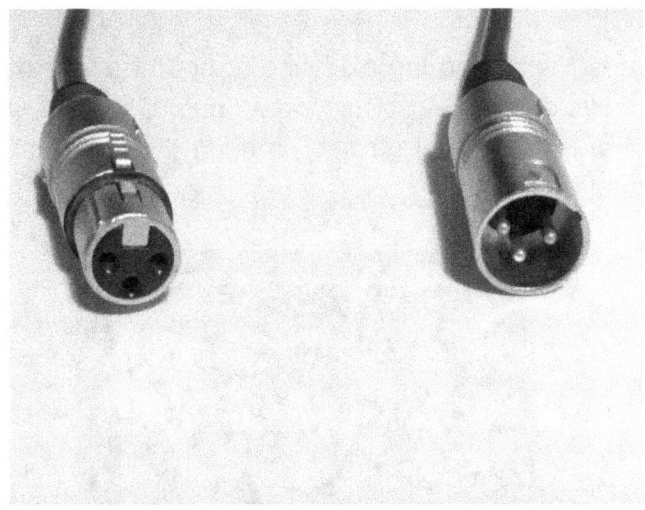

Balanced cables have multiple poles; hot pole (positive polarity), cold pole (negative polarity) and ground.
Balanced cables should be prioritized when available, because they have less chance of receiving external interference such as buzzing, static, hum or power noise.
So, if you need to make a connection where there is a lot of distance between the devices, balanced cables will offer a more powerful signal with less degradation.

To sum up, there are two elements capable of degrading the signal during the passage from a source to the amplifier:

The first, relates to the oxidation of both the terminals of our cable and the electronic input / output sockets, which must always be kept efficient through the use of special products.

The second concerns the radiofrequency signals, present in the environment in which we find ourselves, which can mix with the audio signal during the passage in the cable, up to alter the sound in an audible way.

Regarding digital connections, we can use two types of cable: **coaxial** or **optical**.

Coaxial and optical cables, are used to make audio connections between a source (cd player, cd recorder, MiniDisc or DAT) and an amplifier, or in the case of a source divided into two frames, such as between cd player and dac.

Coaxial digital audio cable

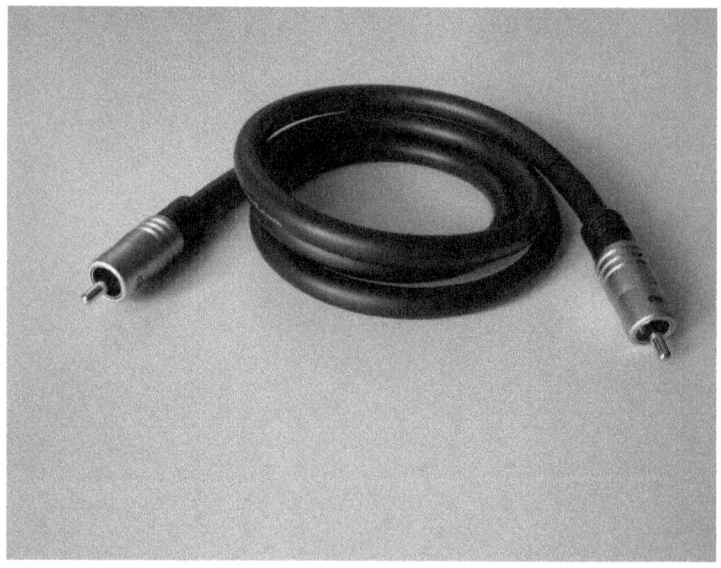

Unlike a traditional RCA cable, like the one on page 113 (which is a double wire, since it carries the signal of the right channel and the left channel separately), the <u>coaxial cable</u> is a single wire and always uses RCA jacks, which are reliable and remain firmly connected.

Coaxial cables, looks apparently identical to a single RCA cable but they're specially engineered to handle the wide bandwidth of digital signals.

Since it's sensitive to radio frequency and electromagnetic interference, I recommend the purchase of a good shielded cable and possibly as short as possible (1 meter).

This is because coaxial cables are known for signal attenuation over long distances, which however should not be a problem for the average home user; if the distance is a problem, buying an optical digital cable, is the better choice.

The underlined optical digital cable (also known as Toslink), transfers audio signals thorough a beam of light that is projected through a plastic or glass fiber (in the best quality cables).

Optical digital audio cable

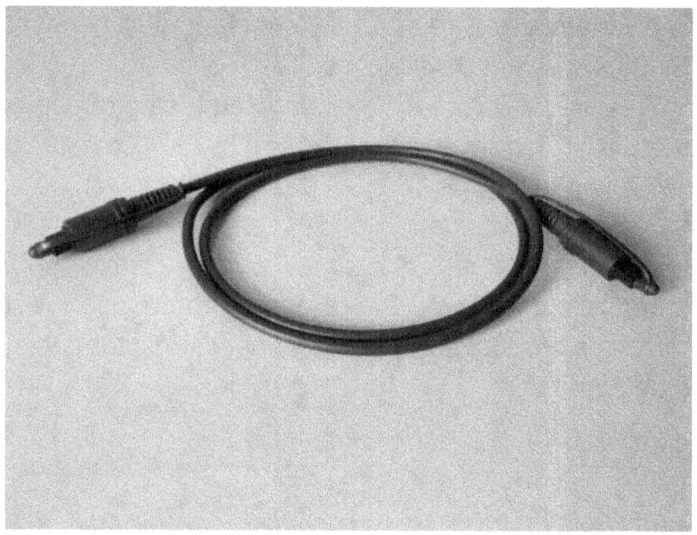

The signal that travels through the cable form the source, must first be converted from an electrical signal to an optical one; when the signal reaches the receiver, it undergoes another conversion from optical to electric. These conversions add digital distortion (jitter).

Optical digital cable connector

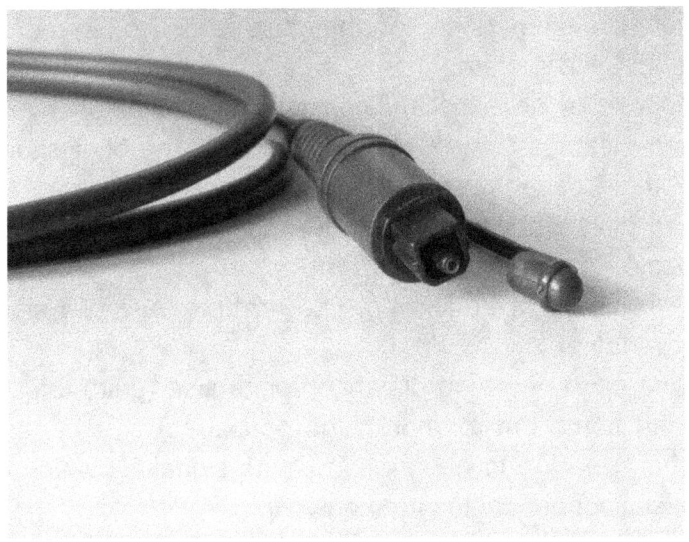

Unlike coaxial cable, the optical one is not sensitive to radio frequency and electromagnetic interference and is not subject to signal attenuation over long distances, that occurs in copper cables.
Optical cables are more fragile than coaxial cables, and cannot be crusched or bent tightly.

Which to choose: optical or coaxial?

There are differences and they can be noticed especially on high-end systems.

The use of an excellent coaxial cable shielded over a short distance (within 1 meter) is certainly the best choice, not really based on the type of cable (slightly better, on a short distance, compared to the optical one), but for the double conversion that it suffers the optical signal with the consequent addition of jitter errors causing a signal degradation.

In the case of a long distance connection (quite unusual in the home environment), the optical cable is certainly better, due to the attenuation of the signal for which the coaxial copper cables are known to.

Let's move on to the *"power" cables*, which are the ones that connect the amplifier output terminals to the loudspeakers input terminals carrying electrical current between the amplifier and the speakers.

There are many kind of power cables, with different sections and connectors.

Many vintage amplifiers accept only one millimeter section or slightly more stripped cable, while the more modern and powerful ones accept large section cable with banana connectors.

Stripped wire, without connectors

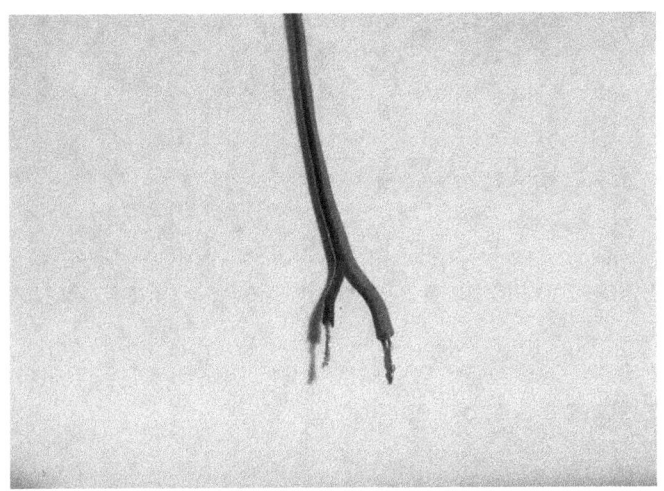

Power cable with large section and banana connectors (commonly banana plugs).

The cable section will be evaluated based on the power of your amplifier and also on the length of the cable you'll use.
The greater the cable lenght and the power of the amplifier, the larger the cable section will be.

The wrong choice of the section of a power cable (test carried out on the same loudspeaker), will generate a sudden decay of the signal related to high frequencies starting from 6.300 Hz up to 20.000 Hz, due to the lack of capacity by the amplifier, to manage the load required by the impedance of the loudspeakers, connected through a power cable with too thin section.
A better quality loudspeaker cable will improve everything you listen to.

Different section types compared.

Let's see how to select <u>the right gauge (thickness).</u>
The lower the gauge number, larger is the cable.
Larger cable such as 12 or 14 gauge (2,05 or 1,63 millimeters) is recommended for long wire distances, high power amplifiers, and low impedance loudspeakers (4/6 ohms).
For short distances, less than 50 feet (15 meters or so) to 8 ohms loudspeakers a 16 gauge (1,29 millimeters) cable will be ok; as example, up to 16 feet (5 meters) this kind of thickness will be ok for an amplifier up to 90/100 watts rms per channel hooked up to loudspeakers with 6/8 ohms impedance.

Power cable with fork termination.

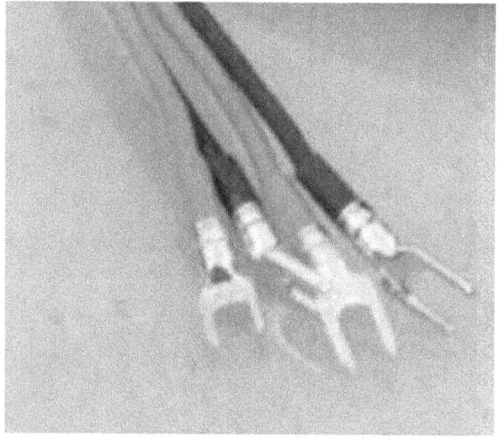

To check out how much cable you need, you can run a cord from your amplifier to each one of the loudspeakers.

IMPORTANT: Although the power cables are less subject to interference, must be as short as possible, both of the same lenght and in any case not longer than 10/16 feet (3/5 meters) between the amplifier and each loudspeaker.

Chapter VI
Listening room and sound system positioning

Now that after many sacrifices divided over time, money spent, choice of components and lots of patience, we have finally arrived at the positioning and "tuning" of our listening room.

As mentioned earlier, in fact, you can have the system with the best and most expensive components of the world, but if you don't position them in the environment in order to express themselves in the best way, all the effort and the money spent so far will be useless.

Each room has its acoustic characteristics, just think about the cathedral acoustics rather than a cinema or a theater.

In fact, sound waves are absorbed in a different way depending on the characteristics of the environment in which they are propagated.

Two very important factors for a listening room, are <u>reverberation</u> and <u>background noise</u>.

The reverberation indicates the persistence of a sound within an environment, after the source has ceased irradiation. Inside a large cathedral, we'll hear an abundant reverberation due to the poor ability to absorb acoustic energy; for example, a voice or the performance of a piece on the piano become very difficult to listen / understand, precisely because of the excessive reverberation.

Regarding the listening room, all the elements present, people, objects, furnishings, walls, windows, influence the acoustic propagation.
The higher the absorption, the shorter the reverberation time.

When instead we go to the cinema or theater, thanks to the excellent acoustics of the environment, we can immediately identify the gentlemen who coughed in the fourth row of the cinema, despite the presence of more than one hundred people in the room (for those who play an instrument instead, refers to the ease of hearing one's own instrument compared to the orchestra on stage); this is because the environmental noise is very low (almost zero); the floors, walls and ceiling are acoustically treated to prevent ambient noise from joining the sound of the instruments / audio system, generating a reduction in sound clarity / timbre variations.

Regarding the room you'll dedicate to listening, it's important that it's not empty; this is to avoid that the sound is reflected from all sides (left, right, front, bottom, floor and ceiling) like a pinball ball.

Starting from the floor, fitted-carpet is an excellent material for sound absorption; alternatively you can use very large rugs to cover most of the floor. Even the upholstered sofas, the velvet curtains (or other heavy fabric), the furniture and paintings (without the glass) on the walls result in good sound absorption, as well as the presence of people in the room; all this helps to contain and reduce reverberation time.

Regarding the sound performances of a system, <u>the correct positioning of the loudspeakers</u> is essential.

If the loudspeakers were positioned randomly, the involvement effect that is the result of an excellent high fidelity system, would be lost.

It would be like looking at the garden from your home window during a beautiful day, instead of sitting on a deck chair sunbathing in the garden.

In the instructions of the best loudspeakers, find the manufacturer's indications with the suggested optimal positioning.

However, there are rules to be followed so that the sounds reach our ears properly.

In fact, the only case in which the sounds reach our ears directly is when we wear a pair of headphones.

Using the loudspeakers instead, you need to orient them so that the sound reaches the listening point correctly.

It should also be said that since not all listening environments are the same in terms of size, it will still be necessary to carry out positioning tests in the listening environment.

If the correct positioning hinders the passage in the room, you can put adhesive markers on the floor, so as to position the loudspeakers correctly every time you listen to your system.

In case you have very large and heavy floor speakers, place them on a platform with wheels that has the same dimensions as the base of the loudspeaker, so that you can easily move them when you're going to do a listening session.

Platform with wheels

Regarding the distance from the walls, remember to position them at least 20" (50 centimeters) from the side walls and the back wall.

Tweeter positioning (the speaker the reproduces the highest frequencies) of the loudspeakers is fundmental; in fact, for otpimal listening, it must be positioned at the same height of the hears of the listener sitting opposite.

With floor loudspeakers it's assumed that all the speakers are mounted at the right distance from the ground and that the tweeter is at the height of the ears of the listener sitting opposite.

Some studio monitors from the past, on the other hand, being designed to be installed above the technicians' seat, were equipped with fins oriented downwards in order to direct the high frequencies of the tweeter towards the ears of the technicians.

Recording studio console and fins detail

Bookshelf loudspeakers, on the other hand, although they are called this way because of the dimensions suitable for being placed on a shelf, shouldn't absolutely be positioned on the bookshelf until after the end of the listening session.

Being smaller than the floor loudspeakers, they need to be positioned above the stands, so that the tweeter is at the height of the ears of the listener sitting opposite.

If you can't finde a pair of the right size, you can build them by yourself if you love DIY, or you can ask to a good carpenter to do it.

Loudspeaker stand

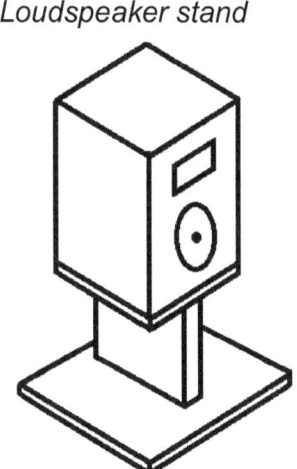

Trough the stand, it's possible to position the tweeter of a bookshelf loudspeaker at the correct height from the ground, like a floor loudspeaker.

Floorstanding and bookshelf loudspeakers

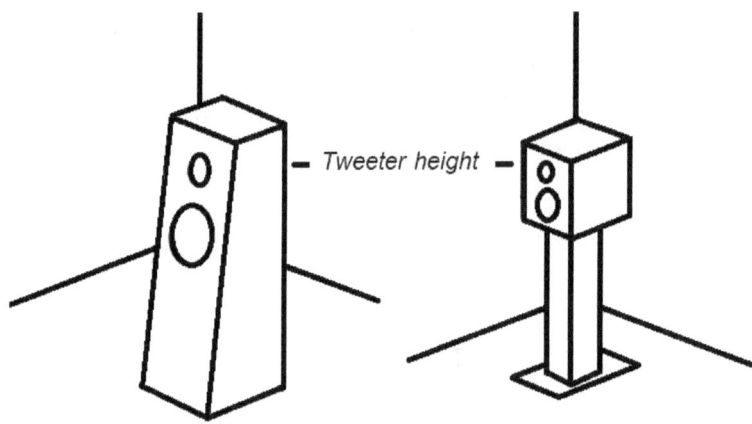

Let's move on to the positioning of the loudspeakers in the listening room.

The ideal listening point should be at the top of a triangle and with the loudspeakers distant from each other, but also from the side and bottom walls, as indicated according to Cardas theory:

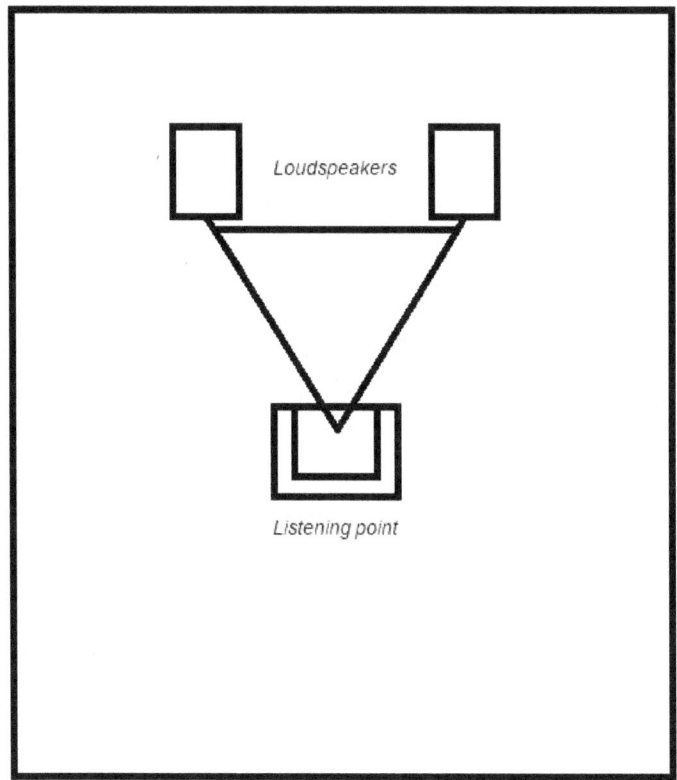

Theory of Cardas

If you use horn loudspeakers, to make them sound at their best, you'll have to place them in the corners of the room as in the figure below:

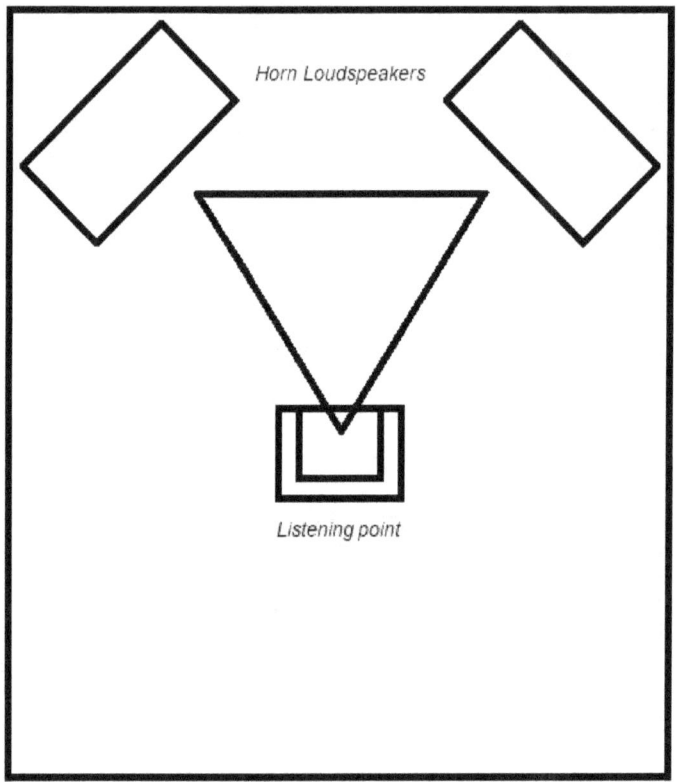

In fact, in order to give maximum performances, the horn loudspeakers need to be positioned in the corners of the room as shown in the above figure .

Here is a positioning according to Wilson also called W.A.S.P.(Wilson Audio Setup Procedure).

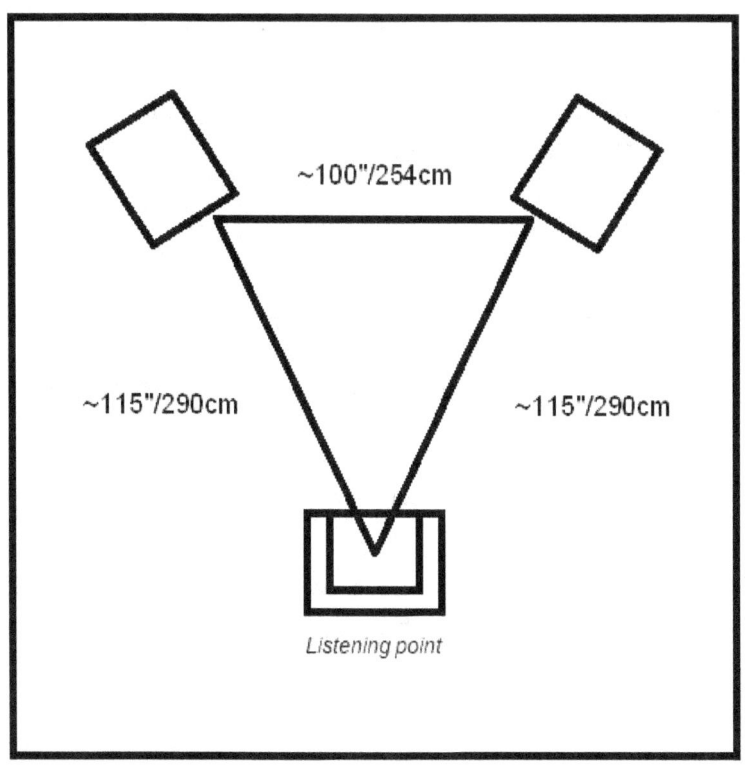

Theory of Wilson

The <u>most suitable distance between one loudspeaker and another</u>, depends on the kind of music you listen to.

If you mainly listen to music played by large orchestras, it's convenient to increase the distance between the loudspeakers and bring the listening point closer (1), for the correct return of the sound scene and to be able to better identify the positions of the various instruments on the stage.

If, on the other hand, you listen to a solo singer, who plays a guitar or piano, the loudspeakers can be positioned one and a half meters from each other or even less and the rearmost listening point (2).

The closer the loudspeakers get, the more the stereophonic effect decreases; however, when listening to a single instrument the stereophonic effect decreases.

In example (1), where the loudspeakers are 3 meters apart and the listening point 1,5 meters from the loudspeakers, 16 positions are recognizable.

In example (2), where the loudspeakers are 2 meters apart and the listening point 3 meters from the loudspeakers, 10 positions are recognizable.

In summary, the distance between the loudspeakers determines the size of the sound scene and the separation between the various musical instruments and the sounds that compose it.

The greater the distance between the loudspeakers, with the listening point close (1), the wider the sound front will be; conversely, the closer the loudspeakers are and with the rearmost listening point (2), the lower the stereophonic effect with a significant reduction in the sound scene.

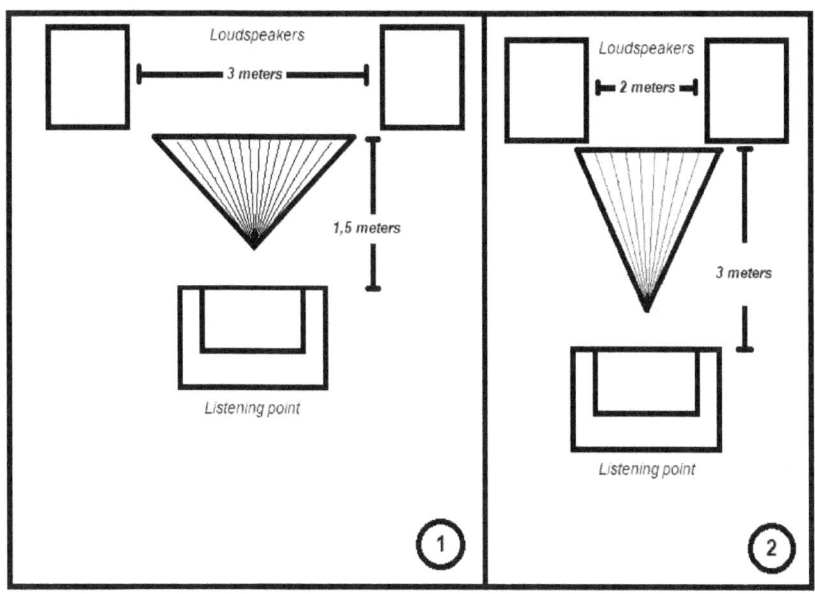

To better calibrate the positioning of the loudspeakers, listen to a song with a solo voice and rotate a few degrees inwards, pointing towards the listening point, until the voice not seem to come exactly from the central point between the two loudspeakers; be careful not to converge the loudspeakers too much, because this will compromise the natural depth of the sound scene.
Do some tests, trying to find the right balance between the focus of the imagine and the depth of the sound scene.

Another important factor that will influence the performances of the sound system is the <u>positioning of the listening point.</u>
The best place is at two thirds of the length of the room.
Unfortunately in many homes the listening point is close to the back wall and this causes an annoying reinforcement of low frequencies.
The distance of both tweeters (left and right) <u>must be always equal</u> from the listening point.

Before proceeding with the purchase of the components of the audio chain, draw the floor plan of the room that you'll dedicate to listening and take the measurements well.

In this way, you'll avoid positioning and space problems due to the furniture already present and you'll avoid being unable to make connections or solve other problems.

Copyright © 2020 Riccardo Ruggiu
All rights reserved

RICCARDO RUGGIU was born in Cagliari in 1976.
He works in the silicon-based technology sector from over twenty years, and developed a deep experience working for the largest and major IT companies in Italy and abroad.
Music has a great impact in his spare time; he is an expert musician (piano, synth and guitar), Dj and deep connoisseur of home and professional audio equipment.

After publishing "Stupidario tecnico: 101 frasi dette dai clienti all'Help Desk" and "How to look for and get a job", has devoted himself to writing this manual for the choice and set up of the Hi-fi system.

Riccardo would love to hear about your experiences with this book (the good, the bad, and the ugly).
You can write to him at: howtobuyhighfidelity@gmail.com.